Episodes in the Mathematics of Medieval Islam

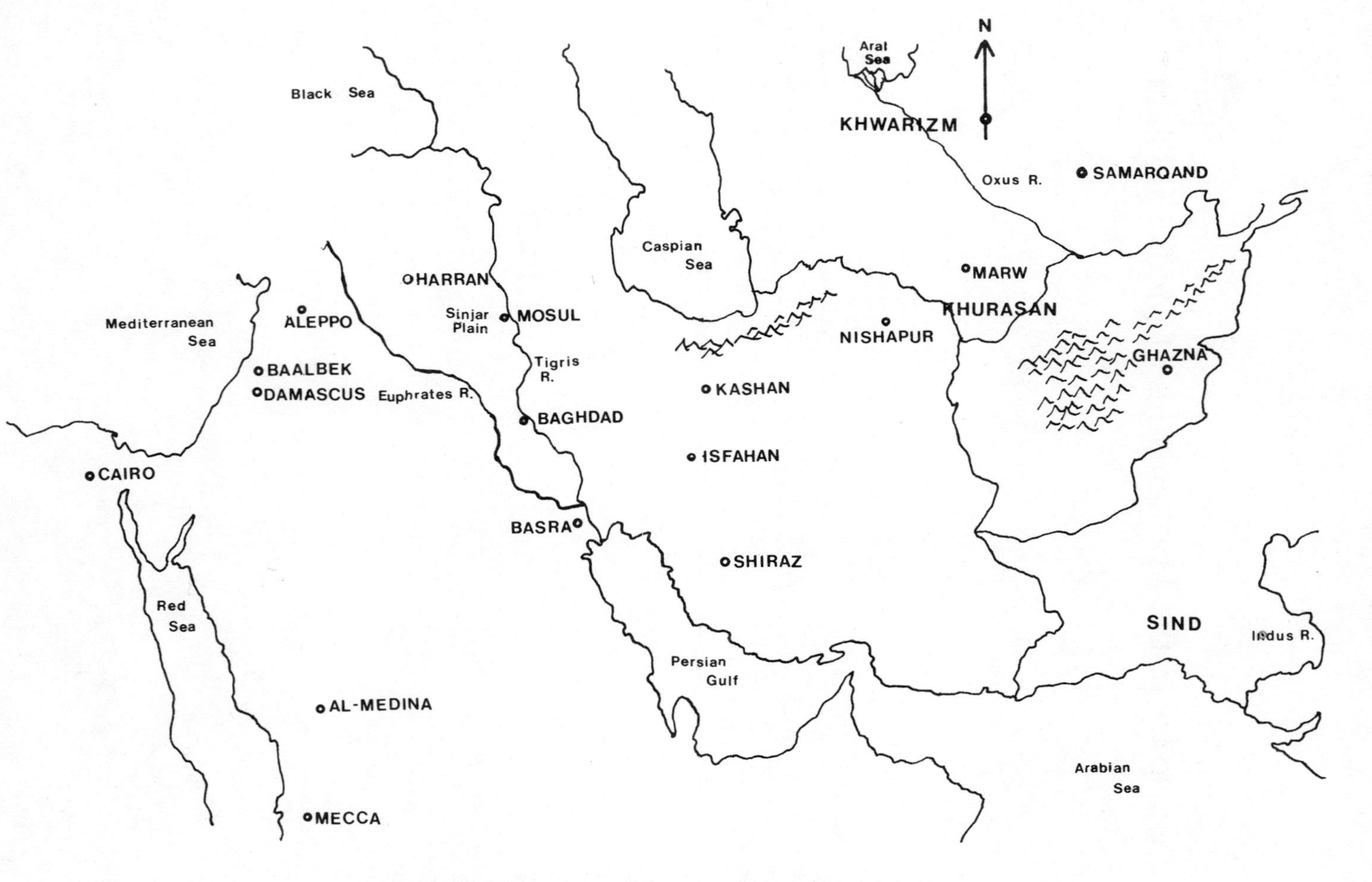

MAP SHOWING MAJOR CITIES IN TEXT

J. L. Berggren

Episodes in the Mathematics of Medieval Islam

With 97 Figures and 20 Plates

Springer-Verlag
New York Berlin Heidelberg
London Paris Tokyo

J. L. Berggren
Department of Mathematics and Statistics
Simon Fraser University
Burnaby, British Columbia V5A 1S6
Canada

AMS Classifications: 01-A05, 01A30

Library of Congress Cataloging in Publication Data
Berggren, J. L.
Episodes in the mathematics of medieval Islam.
Includes bibliographies and index.
1. Mathematics, Arabic. 2. Mathematics—Arab countries—History. I. Title.
QA27.A67B46 1986 510'.917'4927 86-6546

Typeset by Asco Trade Typesetting Ltd., Hong Kong.
Printed and bound by R. R. Donnelley & Sons, Harrisonburg, Virginia.
Printed in the United States of America.

9 8 7 6 5 4 3 2 1

ISBN 0-387-96318-9 Springer-Verlag New York Berlin Heidelberg
ISBN 3-540-96318-9 Springer-Verlag Berlin Heidelberg New York

To My Parents
Evelyn and Thorsten

Preface

Many people today know something of the debt our mathematics owes medieval Islamic civilization. They know "algebra" is an Arabic word and they speak of our Arabic numerals. In recent years, as well, historians of mathematics have re-learned what our medieval and Renaissance forbears knew: the Islamic contribution affected the development of all branches of mathematics in the West and was of prime importance in many. Despite this, no textbook on the history of mathematics in English deals with the Islamic contribution in more than a general way. This is unfortunate, not only from a scholarly point of view but from a pedagogical one as well, for Islam's contributions include some gems of mathematical reasoning, accessible to anyone who has learned high school mathematics. Many of these mark important stages in the development of decimal arithmetic, plane and spherical trigonometry, algebra and such mathematical methods as interpolation and approximation of roots of equations.

The present text is an attempt to fill such a need. My original idea was to write a book modelled on that of the admirable *Episodes from the Early History of Mathematics* by A. Aaboe. However, I soon realized that for the history of mathematics in the world of Islam there was no general backdrop, against which I could highlight certain episodes, such as T. L. Heath's two-volume *History of Greek Mathematics* provides for the Greek world. In addition, the Islamic material contains a relatively larger number of short but important treatises than does the Greek, where a few major, polished works dominate the landscape. For both these reasons, then, I have had to fill in more background and provide a more connected account of 600 years of activity than I first intended. Despite this, the book reflects my conviction that a proper study of the history of mathematics begins with a study of the texts themselves, and for this reason I have written the chapters not as catalogues of results but rather as expositions of portions of mathematical treatises, which I have followed as closely as I feel is possible for an introductory study. When I have departed from the texts it has most often been in the use of abbreviations or symbols where the texts employ words. In any case I have always tried to make the extent of my departures clear enough so that the reader could form a notion of the character of the treatise in question.

Although the study of the history of mathematics begins with a reading of the texts one soon becomes aware that these texts have a context. Each forms a segment of a network of treatises embedded in a particular civilization, and this civilization is itself related, geographically and historically, to other civilizations. I have to some extent tried to exhibit these connections. Also, my study of Islamic mathematics has made me aware that some features of it were responses to the evolving requirements of Islamic faith, and I have tried to indicate some of these features in sections headed "The Islamic Dimension" in Chapters 1, 2, 4 and 6. The section on Islamic art in Chapter 3 also bears on this matter.

At the same time Islamic mathematics, like the civilization itself, was deeply influenced by other great civilizations, some prior to it and others contemporary with it. For the history of mathematics the most important of these were the Greek and Hindu civilizations. Accordingly, I have tried to point out those parts of Greek and Hindu mathematics that the mathematicians of the Islamic world used.

I have not attempted to write "The History of Mathematics in Medieval Islam". Such a book could not be written yet, for so much material remains unstudied that we do not know enough of the whole story. (The nearest we have to such a work is A. P. Youschkevitch's excellent *Les Mathématiques Arabes* (*VIIe–XVe Siècles*). Paris: J. Vrin, 1976.) My aim, rather, has been to exhibit some ways in which writers of the Islamic world contributed to the development of the mathematics one learns in high school, and so the principal mathematical topics treated in this book are arithmetic, algebra, geometry and trigonometry. Even these I have not treated exhaustively, and in particular I have avoided some aspects of them that go well beyond high school mathematics. Also, I have concentrated on developments in the Eastern parts of the Islamic world, largely because this is the area I know best and it proved possible to illustrate the points I wanted to make with examples chosen from this area.

In this book I have restricted the word "Arabic" to designate only the language, and I describe as "Arab" only one coming from the Arabian peninsula. Many who could properly consider themselves to be Arabs are excluded on the basis of my usage but the meaning of "Arab", even within the Arabic-speaking world, has shifted too much over the centuries for the word to be of much use to me. I prefer the designation "Islamic" for that civilization whose mathematical achievements I shall describe. For, although it was home to men and women of many different races and faiths, its essential features were defined by those who professed the Islamic belief that "There is no god but God and Muḥammad is the Messenger of God".

There are a few features of the book which should be mentioned here. I have included a map (page ii) so that the reader may locate the places in which so much of the story is set. I have also included photographs of places and art germane to my account in order, for example, that seeing al-Kāshī's observatory may help fix his name in the reader's mind.

Lest the reader be confused by dates of the form "946–947" I should mention here that the Muslim year is a lunar, not a solar, year. In consequence it is about 11 days shorter than the Western year and it frequently happens that a Muslim year begins in one Western year and ends in the next. Hence, when our Arabic sources tell us only that Abū Fulān was born in a certain Muslim year we cannot usually be more specific than assigning two possible Western years. Finally, a reference in the text of the form "Smith" (or Smith, 1984) refers to the work by Smith listed in the bibliography at the end of the chapter (or to the work he published in 1984 in case more than one work of his is listed).

It remains to thank those who have helped me in the preparation of this book. Asger Aaboe of Yale University has shaped much of my approach to the study of the history of mathematics, and for his years of instruction, encouragement and friendship I offer my thanks. E. S. Kennedy provided me access to his remarkable library in ʿAinab, Lebanon when I began reading for this book, and he and his wife Mary Helen offered my family abundant hospitality, both in Lebanon and Syria. To both of them warm thanks are due. I also wish to thank H. E. Kassis and D. A. King for making photographs available to me from their private collections. I say "Tack så mycket" to Arne Broman and Jöran Friberg, of Chalmers Tekniska Högskola in Göteborg, Sweden, who arranged for me to give the first version of this book in the form of lectures at the Mathematics Department there, and to the many faculty and students at Chalmers who showed such an interest in these lectures. I also thank C. Anagnostakis, J. Hogendijk and W. B. Mc Dermott, all of whom made detailed comments on parts of an early draft of this book, and G. van Brummelen for help reading proofs.

Finally, I want to thank my wife Tasoula who has shared my enthusiasm for the history of mathematics, and our sons, Thorsten and Karl, who offered so many hours of help on illustrations and corrections to the text.

Contents

Chapter 1
Introduction

§1. The Beginnings of Islam

The Muslim calendar begins in the year A.D. 622, when Muḥammad fled from his hometown of Mecca, on the west coast of the Arabian peninsula, to Medīna, a city about 200 miles to the north. The doctrines of one God, called in Arabic *Allāh* = The God, which he announced had been revealed to him by the angel Gabriel, had created considerable dissension in Mecca. This was because Mecca was at that time a thriving center of pilgrimage whose chief attraction was a shrine called the Kaʿba, dedicated to the worship of many gods. Eight years later Muḥammad returned in triumph to Mecca, an event which marked the beginning of the spread of the religion of Islam, based on the idea of *submission* to the will of God, which is the meaning of the Arabic word *Islām*.

When Muḥammad returned to Mecca in A.D. 630, and even when he died in 632, Islamic contributions to the sciences lay in the future. The first "worlds to conquer" were not intellectual but the actual lands beyond the Arabian peninsula, and the Muslims proved as successful in these physical conquests as they were to prove later in the intellectual ones. To give a detailed account of the great battles and the generals whose tactics carried so many days is not our purpose here, and we can only mention that it was Syria and Iraq to the north of the peninsula, whose foreign rulers were heartily detested by the local populations, that fell soonest to the Arab armies and their victory cry "Allāhu akbar" (God is greatest). By 642 the conquest even of Persia was complete, and Islam had reached the borders of India. A few years before this, the general ʿAmr ibn al-ʿAṣ had conquered first Egypt and then all of North Africa, driving the Byzantine armies before him. Soon the new religion had spread from the borders of China to Spain, only to be stopped in France by the victory of Charles Martel near Tours in 732. Although much has been made of this battle, it is relevant to recall the words of an eminent historian, Phillip Hitti, in *The Arabs: A Short History*, where he says:

> Later legends embellished this day . . . , greatly exaggerating its importance. To the Christians it meant the turning point in the military fortunes of their eternal foe. Gibbon, and after him other historians, would see mosques in Paris and London, where cathedrals now stand, and would hear the Koran instead of the Bible expounded in Oxford and other seats of learning, had the Arabs won the day. To several modern historical writers this battle of Tours is one of the decisive battles in history.
>
> In reality the battle of Tours decided nothing. The Arab–Berber wave, already almost a thousand miles from its starting place in Gibraltar, had reached a natural standstill. It had lost its momentum and spent itself. Although this defeat near Tours was not the actual cause of the Arab halt, it does set the farthest limit of the victorious Moslem arms.

The political center of the great empire of early Islam was Damascus, a city so beautiful that Muḥammad, when he saw it, turned back saying that he wanted to enter Paradise only once. Here the caliphs, the succesors of Muḥammad as a political and military leader, held court. They were members of the Umayyad family (see Plate 1.1), but in 750 power passed to a new family known as the ʿAbbāsids, whose power-base was in the Eastern lands.

§2. Islam's Reception of Foreign Science

With this slight account of the early military and political history of Islam we may turn to the beginnings of scientific activity in that civilization, for as early as the time of the Umayyads in the 730's in Sind (modern Pakistan) and Afghanistan astronomical treatises, based on Indian and Persian sources, were written in Arabic. During the time of the ʿAbbāsid caliph, al-Manṣūr, a delegation from Sind which arrived in Baghdad included an Indian who was versed in astronomy and who helped al-Fazārī translate a Sanskrit astronomical text. The resulting work is the *Zīj al-Sindhind*, which contains elements of many astronomical traditions, including mathematical methods using sines.

The caliph al-Manṣūr, who had Baghdad built as his new capital, ordered that the work commence at a time on 30 July 762 that his astrologers considered to be auspicious. (One of these astrologers was al-Fazārī, mentioned above.) The astrologers must have done their work well, for Baghdad did indeed flourish, both as a commercial and intellectual center. Thus, during the reign of Hārūn al-Rashīd, whose glittering rule (786–809) is portrayed in the tales of *The Thousand and One Nights*, a library was constructed, and hére one could doubtless find both originals and translations of scientific works in Sanskrit, Persian and Greek whose contents inspired and instructed the first Islamic scientists. Even more stimulus was given to scientific

Plate 1.1. The Umayyad mosque in Damascus where timekeepers (*muwaqqits*) such as Ibn al-Shāṭir and al-Khalīlī worked. From the minaret the faithful are called to prayer five times a day, which times are astronomically defined. There is the additional requirement that the one saying the prayer faces in the direction of Mecca. and the functioning of such a mosque depended on a certain amount of nontrivial mathematics and astronomy.

activity by the Caliph al-Ma'mūn, who reigned during 813–833, when he founded a translation and research institute known as "The House of Wisdom". In this one institute in Baghdad were housed the translators, as well as the accompanying staff of copyists and binders, who were making the scientific classics of Greek, Syriac, Pahlavi and Sanskrit available to readers of Arabic. These translators, who were sometimes patronized by wealthy families as well as by the caliph, exploited the potential inherent in the Arabic language of expressing subtle variations on an idea by using a set of more or less standard variations of a basic, root form of a word. They thereby created what was to become the language of scholars from North Africa to the borders of China. In Baghdad there was also located an observatory, and numbered among the staff of both the observatory and the House of Wisdom were some of the greatest scholars of their time.

Additionally, the early caliphs sent missions to foreign lands to obtain copies of important books to translate. A good example of the dedicated search for foreign books is the troubles that one of the ninth-century translators, Ḥunayn ibn Isḥāq, had to find a copy of a book by a Greek medical writer named Galen. (The word "ibn" in the middle of Ḥunayn's name is Arabic for "son of", and in the sequel we shall follow the modern custom of abbreviating it by the single letter "b.".) Ḥunayn says first that his colleague, Gabriel, went to great troubles to find it, and then. according to the translation in Rosenthal, "I myself searched with great zeal in quest of this book over Mesopotamia, all of Syria, in Palestine and Egypt until I came to Alexandria. I found nothing, except in Damascus, about half of it (Galen's book). But what I found was neither successive chapters, nor complete. However, Gabriel (also) found some chapters of this book, which are not the same as those I found". (Four hundred years later this search for foreign science would be repeated, but this time it would be Europeans travelling in Islamic lands for precious scientific manuscripts.)

Further, as we have indicated above, there was also considerable support for scientific activity by some of Baghdad's wealthy citizens, a good example being the three brothers known in Arabic as the Banū Mūsā (the sons of Moses). Apart from travelling even to the Byzantine world to buy books, and conducting their own researches on mathematics and mechanics, these ninth-century scholars were early patrons of Thābit b. Qurra from Ḥarrān (the modern Turkish Diyar bakir) in northern Mesopotamia. Thābit lived from 836 to 901 and his gift for languages gave Arabic some of its best translations from Greek. He was a member of the sect of star worshippers who called themselves Ṣabians (after a sect sanctioned in Book 2, verse 63 of the Qur'ān) to escape forced conversion to Islam, for as polytheists, their religious beliefs would have been abhorrent to Muslims. According to one account, the Bānū Mūsā discovered Thābit's linguistic talents when they met him as a money changer in Ḥarrān on their travels and brought him back with them to Baghdad to work with them in their researches. In addition to his skills as a translator, Thābit's talent for mathematics increased the store of beautiful results in that science, and his talents as a medical practitioner earned him a place of honor in the caliph's retinue.

Other important translators in this early period of Islamic science were Ḥunayn b. Isḥāq, whose search for manuscripts we have already mentioned, as well as his son Isḥāq b. Ḥunayn, Qusṭā b. Lūqā from the Lebanese town of Baalbek, and al-Ḥajjāj b. Maṭar. Chart 1.1 is a summary showing, for some of the Greek mathematical works that we shall refer to in this book, the names of those who translated them into Arabic and the approximate dates of the translations.

ARABIC TRANSLATIONS FROM GREEK

Author	Title	Translator	Date
Euclid	*The Elements*	Al-Ḥajjāj b. Maṭar	Time of Harun al-Rashid and al-Ma'mūn.
		Isḥāq b. Ḥunayn Thābit b. Qurra	Late ninth century Died in 901.
	The Data *The Optics*	Isḥāq b. Ḥunayn	
Archimedes	*Sphere and Cylinder*	Isḥāq b. Ḥunayn Thābit b. Qurra	Revised a poor early ninth-century translation.
	Measurement of the Circle	Thābit b. Qurra	Used commentary of Eutocios
	Heptagon in the Circle	Thābit b. Qurra	Unknown in Greek
	The Lemmas	Thābit b. Qurra	
Apollonios	*The Conics*	Hilāl al-Ḥimṣī, Aḥmad b. Mūsā, Thābit b. Qurra	
Diophantos	*Arithmetic*	Qusṭā b. Lūqā	Died 912
Menelaos	*Spherica*	Ḥunayn b. Isḥāq	Born 809

The fate of Apollonios' *Conics* is instructive. The Arabic bibliographer al-Nadīm tells us that "... after the book was studied it was lost track of until Eutocios of Ascalon made a thorough study of geometry.... After he had collected as much of this volume (*Conics*) as he could he corrected four of its sections. The Banū Mūsā, however, said that the volume had eight sections, the part of it now extant being seven, with a part of the eighth. Hilāl b. Abī al-Ḥimṣī translated the first four sections with the guidance of Aḥmad b. Mūsā, and Thābit b. Qurra al-Ḥarrānī the last three."

Chart 1.1

§3. Four Muslim Scientists

Introduction

Like any other civilization that of Islam was not unwavering in its support of scientists, and not long after the time of al-Ma'mūn, support for the House of Wisdom diminished and the institution soon disappeared. In the next century the scholar al-Sijzī, writing from an unnamed locality, complained that where he lived people considered it lawful to kill mathematicians. (Perhaps this was because most mathematicians were also astronomers, and hence astrologers.) However, whatever hardships the vagaries of a particular ruler might cause in one area were generally compensated for by a generous and enthusiastic patron elsewhere, so that, on the whole, mathematicians and astronomers in Islam could expect both honor and support. For example, the Egyptian ruler al-Ḥākim, of whom we shall say more in Chapter 5, founded a library in 1005 called Dār al-Ḥikma. In addition to providing a reading room and halls for courses of studies, al-Ḥākim paid librarians and ensured that scholars were given pensions to allow them to follow their studies.

Islamic civilization thus produced, from roughly 750 to 1450, a series of mathematicians who have to their credit the completion of the arithmetic of a decimal system that includes decimal fractions, the creation of algebra, important discoveries in plane and spherical trigonometry as well as the systematization of these sciences, and the creation of elegant procedures for finding numerical solutions of equations. This list is by no means exhaustive, and we shall detail not only the above contributions but others as well.

Since the men who made these contributions are probably not well known to the reader, we begin with some biographical material on four whose names will appear repeatedly in the following pages. One is Muḥammad b. Mūsā al-Khwārizmī, who was active in al-Ma'mūn's House of Wisdom. The second is Abu l-Rayḥān al-Bīrūnī, whose long life bridged the tenth and eleventh centuries, and whose learning and creative intellect are still impressive. The third, born shortly before al-Bīrūnī died, is the celebrated ʿUmar al-Khayyāmī, and the fourth, whom a contemporary described as "the pearl of the glory of his age", is Jamshīd al-Kāshī, whose work in Samarqand raised computational mathematics to new heights. Taken together, these men represent the breadth of interest, the depth of investigation, and the height of achievement of the best of the Islamic scholars.

Al-Khwārizmī

The springs which fed Islamic civilization sprang from many lands. Symptomatic of this is the fact that the family of its greatest early scientist, the Central Asian scholar Muḥammad ibn Mūsā al-Khwārizmī, came from the old and high civilization that had grown up in the region of Khwārizm. This is the ancient name for the region around Urgench in the U.S.S.R., a city near the delta of the Amu Dar'ya (Oxus) River on the Aral Sea.

Al-Khwārizmī served the Caliph al-Ma'mūn in the House of Wisdom and is connected to a later caliph, al-Wāthiq (842–847), by the following story told by the historian al-Ṭabarī. It seems that when al-Wāthiq was stricken by a serious illness he asked al-Khwārizmī to tell from his horoscope whether or not he would survive. Al-Khwārizmī assured him he would live another fifty years, but al-Wāthiq died in ten days. Perhaps al-Ṭabarī tells this story to show that even great scientists can make errors, but perhaps he told it as an example of al-Khwārizmī's political astuteness. The hazards of bearing bad news to a king, who might mistake the bearer for the cause, are well known. We shall see in the case of another Khwārizmian, al-Bīrūnī, that he too was very astute politically.

Al-Khwārizmī's principal contributions to the sciences lay in the four areas of arithmetic, algebra, geography and astronomy. In arithmetic and astronomy he introduced Hindu methods to the Islamic world, while his exposition of algebra was of prime importance in the development of that science in Islam. Finally, his achievements in geography earn him a place among the ancient masters of that discipline.

His arithmetical work *The Book of Addition and Subtraction According to the Hindu Calculation* introduced the very useful decimal positional system that the Hindus had developed by the sixth century A.D., along with the ten ciphers which make that system, the one we use today, so convenient. His book was the first Arabic arithmetic to be translated into Latin, and its influence on Western mathematics is illustrated by the derivation of the word *algorithm*. This word is in constant use today in computing science and mathematics to denote any definite procedure for calculating something, and it originated in the corruption of the name al-Khwārizmī to the Latin version *algorismi*.

Al-Khwārizmī's book had an equally important effect on Islamic mathematics, for it provided Islamic mathematicians with a tool that was in constant—though not universal—use from the early ninth century onward. From the oldest surviving Arabic arithmetic, Aḥmad al-Uqlīdisī's *Book of Chapters*, written ca. A.D. 950, to the encyclopedic treatise of 1427 by Jamshīd al-Kāshī, *The Calculators' Key*, decimal arithmetic was an important system of calculation in Islam. By the mid-tenth century Aḥmad b. Ibrāhīm al-Uqlīdisī solved some problems by the use of decimal fractions in his book on Hindu arithmetic, so that in a little over a century al-Khwārizmī's treatise had led to the invention of decimal fractions. These too were used by such Islamic mathematicians as al-Samaw'al ben Yaḥyā al-Maghribī in the twelfth century to find roots of numbers and by al-Kāshī in the fifteenth century to express the ratio of the circumference of a circle to its radius as 6.2831853071795865, a result correct to sixteen decimal places.

Arithmetic was only one area in which al-Khwārizmī made important contributions to Islamic mathematics. His other famous work, written before his *Arithmetic*, is his *Kitāb al-jabr wa l-muqābala* (The Book of Restoring and Balancing), which is dedicated to al-Ma'mūn. This book became the starting point of the subject of algebra for Islamic mathematicians, and it also gave its title to serve as the Western name for the subject, for *algebra* comes from the Arabic *al-jabr*. In this book a variety of influences are evident, including Babylonian and Hindu methods that lead to solutions of what we would call quadratic equations and Greek concerns with classification of problems into different types and geometrical proofs of the validity of the methods involved.

The synthesis of Oriental procedures with Greek proofs is typical of Islam, as is the application of a science to religious law, in this case the thorny problems posed by Islamic inheritance law. A large part of the book is devoted to such problems, and here again al-Khwārizmī's example became the model for later Islamic writers. Thus, after the time of al-Khwārizmī, Abū Kāmil, known as "The Egyptian Reckoner", also wrote on the application of algebra to inheritance problems.

Finally, we must comment on al-Khwārizmī's contribution to the science of cartography. He was part of the team of astronomers employed by al-Ma'mūn to measure the length of one degree along a meridian. Since the time of Aristotle (who wrote in the middle third of the fourth century B.C.), men had known that the earth was spherical and, hence, that multiplication of an accurate value for the length of one degree by 360 would lead to a good estimate for the size of the earth. In the century after Aristotle the scientist Eratosthenes of Alexandria, who was the first scientist to be appointed Librarian of the famous library in that city, used this idea with his knowledge of mathematical astronomy to obtain an estimate of 250,000 stades for the circumference of the earth. This was later shortened by an unknown author to 180,000 stades, a figure far too small but adopted by the astronomer, Klaudios Ptolemaios (Ptolemy) in his *Geography*.

We know that the Hellenistic stade is approximately 600 feet but this was not known to the caliph al-Ma'mūn. As al-Bīrūnī says in his *Coordinates of Cities*, al-Ma'mūn "read in some Greek books that one degree of the meridian is equivalent to 500 stadia.... However, he found that its actual length [i.e. the stade's] was not sufficiently known to the translators to enable them to identify it with local standards of length." Thus al-Ma'mūn ordered a new survey to be made on the large, level plain of Sinjār some 70 miles west of Mosul, and two surveying parties participated. Starting from a common location one party travelled due north and the other due south. In the words of al-Bīrūnī:

> Each party observed the meridian altitude of the sun until they found that the change in its meridian altitude had amounted to one degree, apart from the change due to variation in the declination. While proceeding on their paths, they measured the distances they had traversed, and planted arrows at different stages of their paths (to mark their courses). While on their way back, they verified, by a second survey, their former estimates of the lengths of the courses they had followed, until both parties met at the place whence they had departed. They found that one degree of a terrestrial meridian is equivalent to fifty-six miles. He (Ḥabash) claimed that he had heard Khālid dictating that number to Judge Yaḥyā b. Aktham. So he heard of that achievement from Khālid himself.

Again one sees an Islamic side to this project in the involvement of a jurist, for the law was the Islamic religious law and in this case the jurist (*qāḍī* in Arabic) was the chief justice of Baṣra, Yaḥyā b. Aktham. Al-Bīrūnī goes on to say that a second result was also obtained by the survey, namely $56\frac{2}{3}$

miles/degree, and in fact al-Bīrūnī uses this value in his own computations later on.

Al-Khwārizmī's contribution went beyond this to assist in the construction of a map of the known world, a project that would require solving three problems that combined theory and practice. The first problem was mainly theoretical and required mastery of the methods, such as those explained by Ptolemy in the mid-second century A.D., for mapping a portion of the surface of a sphere (the earth) onto a plane. The second was to use astronomical observations and computations to find the latitude and longitude of important places on the earth's surface. The difficulties involved here are both theoretical and practical. The third problem was to supplement these observations by reports of travellers (always more numerous and usually less reliable than astronomers) on journey-times from one place to another. Among al-Khwārizmī's achievements in his geographical work *The Image of the Earth* were his correction of Ptolemy's exaggerated length of the Mediterranean Sea and his much better description of the geography of Asia and Africa. With such a map the caliph could survey at a glance the extent and shape of the empire he controlled.

Thus it was that on his death al-Khwārizmī's legacy to Islamic society included a way of representing numbers that led to easy methods of computing, even with fractions, a science of algebra that could help settle problems of inheritance, and a map that showed the distribution of cities, seas and islands on the earth's surface.

Al-Bīrūnī

The Central Asian scholar Abu l-Rāyḥan al-Bīrūnī was born in Khwārizm on 4 September, 973. During his youth at least four powers were contending with each other in and around Khwārizm, so that in his early twenties al-Bīrūnī spent much of his time either in hiding or fleeing one king to seek hospitality from another. Despite these setbacks, however, he completed eight works before the age of thirty, including his *Chronology of Ancient Nations*, the sort of work necessary to any astronomer who wanted to use (say) ancient eclipse records and needed to convert the dates given in terms of some exotic calendar into dates in the Muslim calendar. He had also engaged in a famous controversy on the nature of light with a precocious teenager from Bukhara named Abū ʿAlī b. Sinā, known to the West as Avicenna, and he had somehow found the time and the means to construct, and to use, large graduated rings to determine latitudes. Also, in cooperation with Abu l-Wafāʾ in Baghdad, he used an eclipse of the moon as a time signal to determine the longitude difference between Kāth (on the River Oxus) and Baghdad. All these observations and calculations (and those for longitude are indeed difficult) he used in a book called *The Determination of the Coordinates of Localities*, where he continued the tradition of geographical research in Islam that goes back at least to al-Khwārizmī. He records in this book that he wanted to do measurements to settle the discrepancy between the two results he had heard of for the number of miles in a meridian degree, and he writes:

> That difference is a puzzle; it is an incentive for a fresh examination and observations. Who is prepared to help me in this (project)? It requires a strong command over a vast tract of land and extreme caution is needed from the dangerous treacheries of those spread over it. I once chose for this project the localities between Dahistān, in the vicinity of Jurjān and the land of the Ghuzz (Turks), but the findings were not encouraging, and the patrons who financed the project lost interest in it.

He also discovered two new map projections, one known today as the azimuthal equidistant projection and the other as the globular projection. (See Berggren (1982) for details.)

Sometime during his thirtieth year, al-Bīrūnī was able to return to his homeland, where he was patronized by the reigning Shāh Abu l-ʿAbbās Ma'mūn. The shah was pressed on the one side by the local desire for an autonomous kingdom and, on the other, by the clear fact that his kingdom existed at the sufferance of Sultan Maḥmūd of Ghazna (in present-day Afghanistan), and he was glad to use the skillful tongue of al-Bīrūnī to mediate the disputes that arose. Of this al-Bīrūnī wrote perceptively "I was compelled to participate in worldly affairs, which excited the envy of fools, but made the wise pity me."

In 1019 even al-Bīrūnī's "tongue of silver and gold" could no longer control the local situation, and the army killed Shah Ma'mūn. Immediately Sultan Maḥmūd invaded and among the spoils of his conquest he took al-Bīrūnī back to Ghazna as a virtual prisoner. Later on al-Bīrūnī's situation improved and he was able to get astronomical instruments and return to his observations.

Maḥmūd's conquests had already made him master of large parts of India, and al-Bīrūnī went to India where he studied Sanskrit. By asking questions, observing, and reading Sankrit texts he compiled information on all aspects of Indian society and culture. His work *India*, which resulted from this observation and study, is a masterpiece, and is an important source for modern Indologists. His comparisons of Islam with Hinduism are fine examples of comparative religion, and show an honesty that is not often found in treatises on other people's religions. Al-Bīrūnī's treatment of Indian religion contrasts starkly with that of his patron Maḥmūd who carried off valuable booty from Indian temples as well as pieces of a phallic idol, one which he had installed as a foot-scraper at the entrance to a mosque in Ghazna. He finished *India* in 1030, after the death of Sultan Maḥmūd. When the succession was finally settled in favor of Masʿūd, one of Maḥmūd's two sons, al-Bīrūnī lost no time in dedicating a new astronomical work to him,

Plate 1.2. *A gunbad* from Ghazna in Afghanistan, built at the time of Sultan Masʿūd, the eleventh-century patron of the polymath al-Bīrūnī.

the *Masʿūdic Canon*. This may have won certain privileges for him, since he was subsequently allowed to visit his native land again (see Plate 1.2).

Sometime after 1040 he wrote his famous work *Gems*, in which he included results of his experiments on the specific gravity of many valuable stones. Much of this material was used by al-Khāzinī who, in the following century, described the construction and operation of a very accurate hydrostatic balance.

The breadth of al-Bīrūnī's studies, if not already sufficiently demonstrated, is plain from his work *Pharmacology*, which he wrote in his eightieth year while his eyesight and hearing were failing. The bulk of the work is an alphabetical listing of about 720 drugs telling, in addition to the source and therapeutic value of each, common names for the drug in Arabic, Greek, Syriac, Persian, one of the Indian languages, and sometimes in other languages as well.

We close this brief biography of al-Bīrūnī with the words of E. S. Kennedy, whose account in the *Dictionary of Scientific Biography* we have relied on for most of the details above:

> Bīrūnī's interests were very wide and deep, and he labored in almost all the branches of science known in his time. He was not ignorant of philosophy and the speculative disciplines, but his bent was strongly toward the study of observable phenomena, in nature and in man ... about half his total output is in astronomy, astrology, and related subjects, the exact sciences par excellence of those days. Mathematics in its own right came next, but it was invariably applied mathematics.

ʿUmar al-Khayyāmī

ʿUmar al-Khayyāmī must be the only famous mathematician to have had clubs formed in his name. They were not, however, clubs to study his many contributions to science, but to read and discuss the famous verses ascribed to him under the name of *The Rubʿāyāt* (*Quatrains*) which have been translated into so many of the world's languages. Indeed, outside of the Islamic world ʿUmar is admired more as a poet than as a mathematician, and yet his contributions to the sciences of mathematics and astronomy were of the first order.

He was born in Nishāpūr, in a region now part of Iran, but then known as Khurasān, around the year 1048. This is only shortly before the death of al-Bīrūnī, at a time when the Seljuk Turks were masters of Khurasān, a vast region east of what was then Iran, whose principal cities were Nishapur, Balkh, Marw and Ṭūs. His name, "al-Khayyāmī", suggests that either he or his father at one time practiced the trade of tentmaking (*al-khayyām* = tentmaker). In addition, he showed an early interest in the mathematical sciences by writing treatises on arithmetic, algebra and music theory, but, beyond these facts, nothing is known of his youth. The story of a boyhood pact with a schoolmate, who later was known as Niẓām al-Mulk and became a minister in the government of the ruler Malikshah, that whichever of them first obtained high rank would help the other is not supported by the dates at which these men lived. In fact, most scholars believe that ʿUmar died around 1131, so if he had been Niẓām's schoolmate, he would have had to be around 120 years old when he died in order for the story to fit the known dates for Niẓām al-Mulk. Better founded is the report of the biographer Ẓāhir al-Dīn al-Bayhaqī. He knew al-Khayyāmī personally and describes him as being both ill-tempered and narrow-minded. Of course, al-Bayhaqī was examined as a schoolboy by al-Khayyāmī in literature and mathematics, so it may be that he did not get to know him under the best of circumstances.

We also know that in 1070, when ʿUmar wrote his great work on algebra, he was supported by the chief judge of Samarqand, Abū Ṭāhir. In this work, ʿUmar systematically studied all the kinds of cubic equations and used conic sections to construct the roots of these equations as line segments obtained from the intersections of these curves. There is evidence that ʿUmar also tried to find an algebraic formula for these roots, for he wrote that "We have tried to express these roots by algebra but have failed. It may be, however, that men who come after us will succeed." The candor of this passage, and its recognition of being part of a tradition of inquiry that will continue after one's own death, bespeaks, al-Bayhaqī aside, a modest and civilized man.

During the 1070's ʿUmar went to Isfahan, where he stayed for 18 years and, with the support of the ruler Malikshah and his minister Niẓām al-Mulk, conducted a programme of astronomical investigations at an observatory (see Plate 1.3). As a result of these researches, he was able, in 1079, to present a plan to reform the calendar then in use. (One wonders if there is not an echo of this achievement, together with a nice reference to squaring the circle, in the quatrain "Ah, but my calculations, people say,/Have squared the year to human compass, eh?/If so by striking out/Unborn tomorrow and dead yesterday.") ʿUmar's scheme made eight of every 33 years leap years, with 366 days each, and produced a length for the year closer to the true value than does the present-day Gregorian calendar.

Another important work of ʿUmar's was his *Explanation of the Difficulties in the Postulates of Euclid*, a work composed in 1077, two years before he presented his calendar reform. In this treatise, ʿUmar treats two extremely important questions in the foundations of geometry. One of these, already treated by Thābit ibn Qurra and Ibn al-Haytham (known to the West as Alhazen), is the fifth postulate of Book I of Euclid's *Elements* on parallel lines. (In fact, Toth provides evidence that remarks in various writings of Aristotle imply that mathematicians before Euclid investigated this question.) ʿUmar bases his analysis on the quadrilateral ABCD of Fig. 1.1, where CA and DB are two equal line segments, both perpendicular to AB, and he recognizes that in order to show that the parallel postulate follows from the other Euclidean postulates it suffices to show that the interior angles at C and D are both right angles, which implies the existence of a

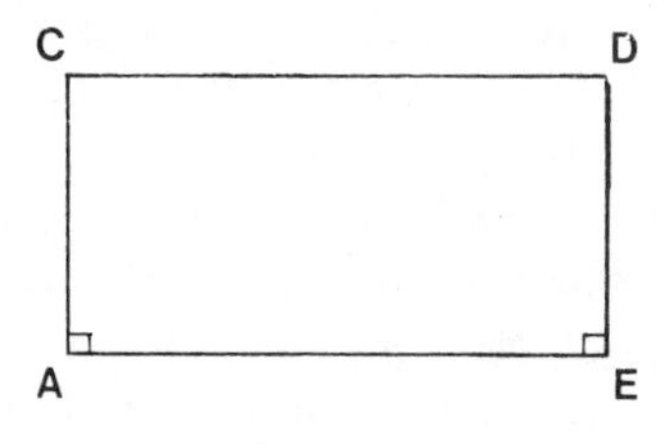

Fig. 1.1

Plate 1.3. The entrance to the Friday Mosque in Isfahan, Friday being the day when Muslims gather together in the mosque to pray and, perhaps, to hear a *khuṭba* (homily). Parts of the mosque date back to the time of ʿUmar al-Khayyām. The tile designs on the twin minarets are calligraphic renderings of the word *Allah* (The One God), and calligraphy borders the geometric patterns of arabesques on the facade.

rectangle. (In fact, the two may be shown to be equivalent.) Although Ibn al-Haytham, who flourished around 1010, preceded ʿUmar in using this method of attack on the problem ʿUmar took issue with Ibn al-Haytham's use of motion in geometry. A century and a half later Naṣīr al-Dīn al-Ṭūsī adopted ʿUmar's view when he wrote his treatment of Euclid's parallel postulate, and Naṣīr al-Dīn al-Ṭūsī's treatise influenced European geometers of the seventeenth and eighteenth centuries.

The other topic that ʿUmar treated in his discussion of the difficulties in Euclid is that of ratios. Here al-Khayyāmī's achievements are twofold. The one is his demonstration that a definition of proportion that was elaborated in Islamic mathematics, a definition that he felt was more true to the intuitive idea of "ratio", was equivalent to the definition Euclid used. The other is his suggestion that the idea of number needed to be enlarged to include a new kind of number, namely ratios of magnitudes. For example, in ʿUmar's view, the ratio of the diagonal of a square to the side ($\sqrt{2}$), or the ratio of the circumference of a circle to its diameter (π), should be considered as new kinds of numbers. This important idea in mathematics amounted to the introduction of positive real numbers and, as was the case with the parallel postulate, this was communicated to European mathematicians through the writings of Naṣīr al-Dīn al-Ṭūsī.

ʿUmar once said to a friend that when he died he wanted to be buried in Isfahan, where "the wind will blow the scent of the roses over my grave." His wish was granted, and the tomb of Islam's poet–mathematician has remained there to this day.

Al-Kāshī

Among the nicknames that were sometimes bestowed on mathematicians or astronomers in the Islamic world was that of *al-Ḥāsib* (=the reckoner). Strangely, however, the man who may be the most deserving of that title seems never to have gotten it, but rather bore the name Ghiyāth al-Dīn Jamshīd al-Kāshī; but, before we speak of his remarkable calculations we must tell what is known of his life.

He was born in the latter half of the fourteenth century in the Persian town of Kāshān, some 90 miles north of ʿUmar's tomb in Isfahan, but we know nothing of his life until the year 1406 when, his own writings say, he began a series of observations of lunar eclipses in Kāshān. In the following year, also in Kāshān, he wrote a work on the dimensions of the cosmos, which he dedicated to a minor prince. Seven years later, in 1414, he finished his revision of the great astronomical tables written 150 years earlier by Naṣīr al-Dīn al-Ṭūsī, and he dedicated this revision to the Great Khan (Khāqān) Ulūgh Beg, the grandson of Tamurlane, whose capital was situated at Samarqand (see Plates 1.4 and 1.5). In the introduction to these tables he speaks of the poverty he endured and how only the generosity of Ulūgh Beg allowed him to complete the work. Then in 1416, two years later, he finished a short work on astronomical instruments in general, dedicated to Sulṭān Iskandar (a Black Sheep Turk and member of a dynasty rival to that of the offspring of Tamurlane), and a longer treatise on an instrument known as an equatorium. This instrument is, in essence, an analog computer

Plate 1.4. The tomb of Tamurlane in Samarqand. Tamurlane was a man of considerable intellectual attainments, as well as being a great military strategist, and his grandson, Ulūgh Beg, was a generous patron of learning and the arts in the first half of the fifteenth century in Samarqand.

for finding the position of the planets according to the geometrical models in Ptolemy's *Almagest*, and its utility is that it allows one to avoid elaborate computations by manipulating a physical model of Ptolemy's theories to find the positions of the planets. See Fig. 1.2 for a picture of al-Kāshī's equatorium.

The writing on the equatorium marked the end of al-Kāshī's career as a wandering scholar, and the next we hear of him is as a member of the entourage of Ulūgh Beg, to whom he had dedicated his Khāqānī tables. Exactly when al-Kāshī arrived in Samarqand we do not know, but during the year 1417 Ulūgh Beg began building a *madrasa* (school) there, whose remains still impress visitors to the site, and, on its completion, began construction on an observatory (see Plates 1.6 and 1.7).

A letter from al-Kāshī to his father has survived and gives us a rare glimpse into details of the intellectual life at Ulūgh Beg's court. In this letter al-Kāshī describes at length the accomplishments of Ulūgh Beg who, he says (translated in Kennedy *et al.*, p. 724)

> "has by heart most of the glorious Qur'ān ... and every day he recites two chapters in the presence of (Qur'ān) memorizers and no mistake is made. He knows (Arabic) grammar well and he writes Arabic composition extremely

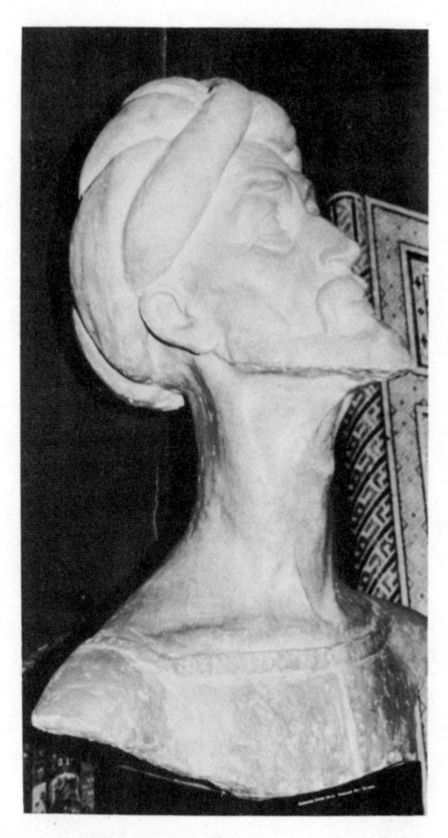

Plate 1.5. A modern bust of al-Kāshī's patron Ulūgh Beg at the museum in Samarqand. Ulūgh Beg was himself an accomplished astronomer whose astronomical tables were used in Europe into the seventeenth century.

> well, and like-wise he is well posted in canon law; he has knowledge of logic, rhetoric, and elocution, and likewise of the Elements (of Euclid?), and he himself cultivates the branches of mathematics, and this has reached the extent that one day while riding he wanted to determine the date, which was a Monday of [the month of] *Rajab*, between the fifth and the tenth in the year eight hundred and eighteen (A.H.), as to what day it was of the (astronomical) season of the year. From these very given data, by mental computation, and from horseback, he determined the true longitude of the sun (correct) to degrees and minutes. When he came back he asked this humble servant about it. Truly, since in mental computation the quantities must be retained by memory and others determined, and there is a limit to one's strength of retention, he (i.e. I) was not able to extract it to degrees and minutes, but contented myself with degrees."

It is perhaps because of Ulūgh Beg's enlightened patronage of learning that al-Kāshī goes on to call Samarqand a place where "the rams of the

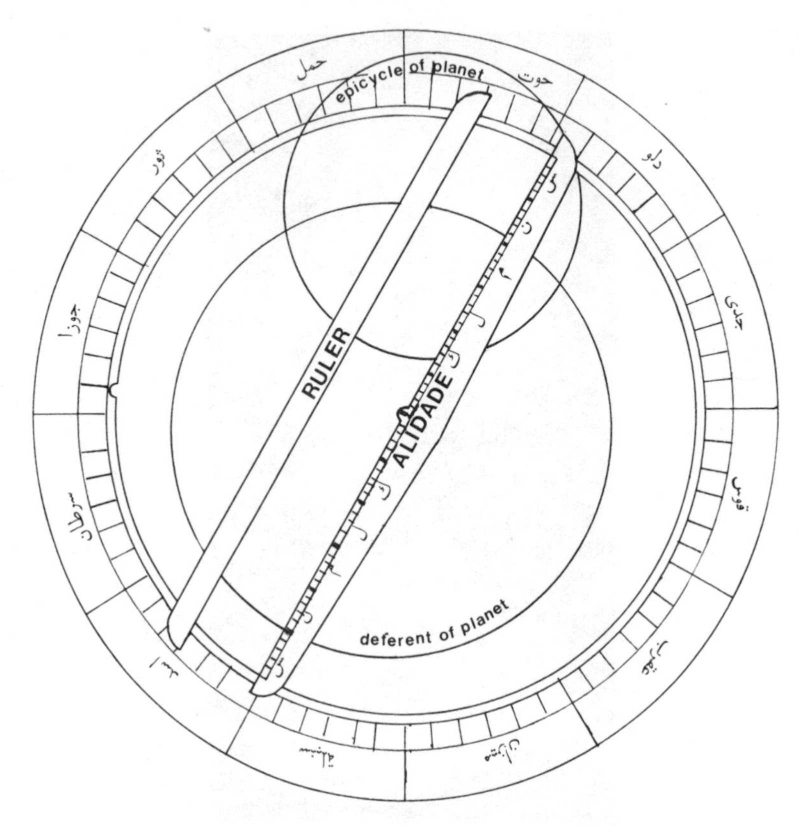

Fig. 1.2. Al-Kāshī's equatorium set for finding the longitude of a planet. The outer rim shows the Arabic names for the zodiacal signs, counterclockwise from Aries, the word just above "epicycle". Figure adapted from *The Planetary Equatorium of Jamshīd Ghiyāth al-Dīn al-Kāshī* (transl. by E. S. Kennedy), Princeton, NJ: Princeton University Press, 1960. Reprinted by permission of Princeton University Press.

learned are gathered together, and teachers who hold classes in all the sciences are at hand, and the students are all at work on the art of mathematics".

However he leaves no doubt in his father's mind that he is the most able of all gathered there. He first tells how he found for the scholars at Samarqand the solution of difficulties involved in laying out the star-map of an astrolabe. He then lists another triumph (Kennedy *et al.*, p. 726):

> Furthermore, it was desired to set up a gnomon on the wall of the royal palace and to draw the lines of the equal hours on it (the wall). Since the wall was neither in the line of the meridian nor in the east–west line, no one had done the like of it (before) and (they) could not do it at all. Some said that it can be done in one year, i.e., as soon as the sun reaches the beginning of a zodiacal sign, on that day let an observation be made for each hour, and a mark made until it is finished. When this humble servant arrived it was commanded that

Plate 1.6. The observatory, as the scientific institution we know today, was born and developed in the Islamic world. Here is part of the sextant (or perhaps quadrant) at the observatory in Samarqand where al-Kāshī worked. It was aligned in the north–south direction and was 11 meters deep at the south end. Thus an astronomer sitting between the guide rails could have seen the stars crossing the meridian even in the daytime while assistants sitting on either side held a sighting plate through which he could observe the transits of heavenly bodies. It was at this observatory that the greatest star catalog since the time of Ptolemy was compiled, and the observatory itself, as the scientific institution we know today, developed in the Islamic world.

> this humble servant (lay out the lines), which was completed in one day. When it was examined by use of a big astrolabe it was found to be in agreement and proper arrangement.

One can imagine what al-Kāshī thought (and probably said) about the persons he mentions who could only draw the lines for a sundial not facing one

Plate 1.7. This mausoleum houses the tomb of al-Kāshī's colleague at Samarqand, Qāḍī Zadeh al-Rūmī, one of the few colleagues he seems to have had some respect for. The building derives its basic structure from a favorite pattern in Muslim architecture—a dome surmounting an octagonal base which is itself resting on a square.

of the cardinal directions by waiting for a year and marking off the shadows month-by-month.

It is from this period in Samarqand, that is from about 1418, that al-Kāshī's greatest mathematical achievements come. One of these, is his spectacular calculation, in 1424, of a value for 2π which, when he expresses it in decimal fractions, is correct to 16 decimal places. In order to achieve this accuracy he calculated the perimeters of inscribed and circumscribed polygons, in a given circle, having 805,306,368 sides. What makes the achievement impressive is that al-Kāshī states in advance how close he wants his approximation to be and then carefully plans how accurate each stage must be so that what we would call round-off errors do not accumulate as he goes through the series of root extractions necessary to arrive at the final result. Al-Kāshī phrases his requirement for accuracy by stating that he wants the value to be so accurate that, when it is used to calculate the circumference of the universe according to the ancient dimensions, the result would not differ from the true value by more than the width of a horse's hair.

Although this treatise on π bears no dedication, the work he completed two years later, a compendium of arithmetic, algebra and measurement called *The Calculators' Key*, is dedicated to Ulūgh Beg, and as the crowning achievement of Islamic arithmetic is a gift fit for a king. Among its many jewels is a systematic exposition of the arithmetic of decimal fractions, an invention al-Kāshī claims as his own, and a beautiful algorithm for finding the fifth root of a number which he demonstrated on a number on the order of trillions. Such was the excellence of the work that, according to the Persian scholar Muḥammad Tāhir Tabarsī, al-Kāshī's book was the standard arithmetic and algebra text in the Persian *madrasa*s until the seventeenth century. It is also of interest that a copy of *The Calculators' Key* found in the British Museum was made by the great-great-great grandson of al-Kāshī.

Finally, among the works al-Kāshī mentions in the preface to *The Calculators' Key* is one on solving a cubic equation to obtain a value for $\mathrm{Sin}(1^\circ)$. A remarkable feature of al-Kāshī's method, whose details we explain in Chapter 5, is that the calculator can repeat the procedure, each time using the last result obtained, and so obtain numbers as near to the true value of the root of the equation as he pleases.

Al-Kāshī's remarkable career ended on the morning of 22 June 1429 when he died at the observatory he had helped to build. In the preface to his own astronomical tables, written some eight years after al-Kāshī's death, Ulūgh Beg refers to al-Kāshī as "the admirable mullah, known among the famous of the world, who had mastered and completed the sciences of the ancients, and who could solve the most difficult questions".

These words are a fitting epitaph, not only for al-Kāshī but for any of the great mathematicians of Islam, and they may end this brief biographical section. (We shall, of course, give biographical details about some of the figures mentioned later in the book as they appear.)

§4. The Sources

Most sources for the history of Islamic mathematics are treatises running from a few pages to several hundred pages, written with ink and usually on paper (see Plate 1.8). Typically, several treatises are bound together to form a

الى سطح م د ز ط وكذلك نبين ان نسبة مثلث ب ه ح الى مثلث ج ز ط
كنسبة سطح ب ه ح ى الى سطح ج ز ط ل لكن نسبة مثلث ا ه ح الى مثلث
د ز ط مثل نسبة مثلث ب ه ح الى مثلث ج ز ط لتساوي د ز ز ج فنسبة
مثلث ا ب ح الى مثلث د ج ط مجموع مقدمين الى مجموع تاليهما
كنسبة شكل ب ا ق ح ى الى شكل ج د م ط ل
مجموع باقي المقدمات الى باقي التوالي لكن نسبة
مثلث ا ب ح الى مثلث د ج ط كانت
كنسبة قطعة ا ح ب الى سطح
ك فنسبة شكل ب ا ق ح
ى الى شكل ج د م ط
ل كنسبة قطعة ا ح
ب الى سطح ك
خلف لان
قطعة ا ح ب
اعظم من شكل
ب ا ق ح ى وسطح
ك قد بينا انه اصغر
من شكل ج د م ط ل
او فليكن نسبة مثلث ا ح ب
الى مثلث د ط ج كنسبة قطعة
ا ح ب الى سطح ك الذي هو اعظم من قطعة
د ط ج فنسبة مثلث د ط ج الى مثلث ا ح ب
كنسبة قطعة د ط ج الى سطح هو اصغر من قطعة ا ح ب
خلف لما بينا فنسبة مثلث ا ح ب الى مثلث د ط ج كنسبة قطعة
ا ح ب الى قطعة د ط ج ده كل قطعة من قطوع مكافئ فهي كثلثي
السطح المتوازي الاضلاع الذي قاعدته قاعدتها وارتفاعه ارتفاعها
وهي مثل وثلث للمثلث الذي قاعدته قاعدتها ورأسه رأسها
فلتكن قطعة ب ا ج من قطوع مكافئ قاعدتها ب ج وننصف ب ج
على د ونخرج من د قطر د ا ونخرج على ا خط ز ا ش يوازي ب ج
وعلى ب وج خطي ب ز ج ش يوازيان د ا ونصل ب ا ا ج فاقول
ان قطعة ب ا ج من القطع ثلثا سطح ب ز ش ج وانها مثل وثلث

volume, called a *codex*, many of which have interesting histories. For example, the owners' marks on the title page of a manuscript now at the library of the great shrine at Meshhed in Iran, which was copied in 1462–1463, show that in the seventeenth century it was in the library of Shāh Jahān in India, the Mughal emperor and patron of the sciences. There are a variety of notations from the late seventeenth century, still at the court of the Mughals. Then, by the nineteenth century the manuscript had reached Meshhed, at a library called the Fāḍilīya, whose contents were recently incorporated into those of the shrine. Some scholars have conjectured that the manuscript came to Meshhed as part of the booty brought back to Iran by Nādir Shāh when he defeated the Mughals in 1739. Certainly it would not be the unique instance of scientific manuscripts being among the spoils of war, for al-Ma'mūn obtained Greek manuscripts from the Byzantines by the terms of peace treaties.

In the previous centuries, as well as in the early part of this one, precious Arabic scientific manuscripts could be obtained very cheaply, on the open market or otherwise. In addition, European collectors could hire an Arabic-writing scribe to copy old manuscripts for them. In this way, through purchase, theft, gift and copying, many large collections were built up by Europeans and these collections were in turn donated to or purchased by European libraries. Thus at Berlin, Dublin, the Escorial in Spain, Leiden, London, Oxford and Paris (to name only some of the main sites) there are large collections of Arabic manuscripts accessible to scholars, and there are also several large collections in both the U.S. and the Soviet Union. (In the latter case, many of the collections are native, since the Central Asian Republics, such as Uzbekistan and Tajikistan, are part of the old Islamic heartlands.)

In the Islamic world there are of course large collections of manuscripts all the way from Morocco through Afghanistan to India and S. E. Asia. Many of these libraries are generous in providing access to the collections, but in other cases local politics and nationalism combine with a lack of adequate catalogs to make access difficult for scholars.

The study of Arabic manuscripts presents the same problem as does the study of Greek manuscripts, namely that one is usually dealing with copies, many times removed, of a vanished original; but, additional difficulties arise because of the fact that, unlike Indo-European languages, written Arabic indicates only consonants and long vowels. Thus, the title of al-Kāshī's book in Arabic is مفتاح الحساب, and the second word, reading in the direction of

---

◁ Plate 1.8. This page of a manuscript housed in the *Dār al-Kutub* ("abode of books") in Cairo is from a treatise by Ibrāhīm b. Sinān on the area of a parabalic section. This copy was made by the scholar Muṣṭafa Sidqī in the early eighteenth century. Perhaps Sidqī's competence in mathematics was one reason the diagrams are so carefully done. (One often finds very nicely written manuscripts with spaces carefully left for diagrams which were never drawn.) Photo courtesy of Egyptian National Library.

Arabic writing from right to left, could be read as *ḥisāb* or as *ḥussāb*, where the first means "arithmetic" and the second "calculators". If the context does not provide clues, and if there are no special marks to indicate which is intended, there is simply no way to tell. A further problem arises because very different letters are distinguished only by one or more dots, and often the dots are omitted. For example, the letters ج ، ح ، خ are, respectively, "J", "H" (approximately), and "Kh", and so, for the famous mathematician الكرجى, depending on whether the ح has a dot written underneath or above, his name will be read "al-Karajī" or "al-Karkhī". In the first case it implies he came from Iran and in the second case from Iraq, but, since scholars say about as many manuscripts support one reading as another, we shall probably never know the origin of one of the greatest Islamic algebraists.

Despite these difficulties, the last few decades have seen an increased interest in the study of all aspects of the mathematical sciences in the Islamic world. And, even though we are far from having the same proportion of carefully edited texts of major works that students of Greek mathematics have at their disposal, we do possess at least the main outlines of the story that will occupy the following chapters.

§5. The Arabic Language and Arabic Names

The Language

The Arabic language is a member of the group of languages known as the Semitic group, a group that includes Hebrew, Ethiopian, Babylonian and Phoenician. All members of this group share the characteristic that most of their words are formed around roots of three consonants. For example, the consonants "k t b" in that order carry the meaning of *writing*. *Kataba* means "he wrote", and if the first and second vowels are changed to "u" and "i" then one has *kutiba*, "it was written".

The roots of Arabic words are quite unfamiliar to us and one learning Arabic must do an immense amount of memorization to learn an entire vocabulary with almost no similarities to English. (One of the few areas where there is any help from English is a few terms in mathematics and astronomy we have borrowed from Arabic.) However, one feature of the Arabic language that does ease the acquisition of its vocabulary is that derived forms of the root are formed in a standard way, independently of the root. To illustrate this consider the three roots *k t b* = writing, *r ṣ d* = observing, *ḥ s b* = computing. If XYZ is any root then XāYiZ denotes a person performing the action of the root, so that *kātib* is a scribe, *rāṣid* a watcher and *ḥāsib* a calculator (or an astronomer, who had to do so many calculations). The place where the action corresponding to the root is done is denoted by the form maXYaZ, so that *maktab* is an office or a desk and *marṣad* is an observatory. As a last example miXYaZ is an instrument for doing something so that *mirṣad* is a telescope. Arabic has standard forms not only for such concrete notions as the above but also for more subtle variations of the basic meaning, and this feature made it possible for the early Arabic translators to find Arabic equivalents for a wide range of concepts in Greek, Persian and Indian science and philosophy.

Transliterating Arabic

The reader perhaps noticed in the examples above that some vowels are written with a bar, which linguists call a macron, over them. In fact the Arabic alphabet has 28 letters, all of them consonants. Short vowels are indicated by the marks "/" (for "a") and "," (for "u") placed above the consonant and "/" for "i" placed under the consonant. Apart from the Muslim scriptures, the Qur'ān, short vowels are written only when the possibility of an ambiguity occurs to the writer. However, since long vowels are used to distinguish one standard derived form from another they must be indicated. This is done in Arabic by using three letters to indicate the three long vowels, "aliph" (ā), "waw" (ū) and "yā'" (ī). It is done in transcription by the macron.

Since English has only 21 consonants to Arabic's 28, some special devices are necessary to transliterate Arabic to English. ("Transliteration" means using one system of representing sounds—e.g. the English alphabet—to represent another—e.g. the Arabic alphabet. For the names in this book we are using the system of transliteration explained in Haywood and Nahmad (except that we do not use any underlining.) This is why the reader will see in transliterated Arabic words not only the consonants h, s, d, t, z but also ḥ, ṣ, ḍ, ṭ, ẓ. Corresponding letters, h and ḥ for example, may be pronounced conveniently, if incorrectly, as identical. (Obviously the sounds must have something in common.) Or the reader may consult Tritton where phonetic equivalents of all the Arabic consonants are given. The only other signs we shall comment on are "ʾ" and "ʿ". The first of these represents "aliph", the first letter of the Arabic alphabet, and is a glottal stop, the first clicking sound we make when emphasizing a word like "am". The second represents the eighteenth letter, "ʿayn", and is pronounced as a sort of short growl, deep in the throat. It has no English equivalent—or even approximant.

Arabic Names

A child of a Muslim family will receive a name (called in Arabic *ism*) like Muḥammad, Ḥusain, Thābit, etc. After this comes the phrase "son of so-and-so", and the child will be known as Thābit ibn Qurra (son of Qurra) or

Muḥammad ibn Ḥusain (son of Ḥusain). The genealogy can be compounded. For example, Ibrāhīm ibn Sinān ibn Thābit ibn Qurra, carries it back to the great-grandfather. Later in life one might have a child and then gain a paternal name (*kunya* in Arabic) such as Abū ʿAbdullāh (the father of ʿAbdullāh). Next comes a name indicating the tribe or place of origin (in Arabic *nisba*), such as al-Ḥarrānī, "the man from Harrān". At the end of the name might come a tag (*laqab* in Arabic), it being a nickname such as "the goggle-eyed" (al-Jāḥiẓ) or "the tent-maker" (al-Khayyāmī) or a title such as "the orthodox" (al-Rashīd) or "the blood-shedder" (al-Saffāḥ). Putting all this together, we find the name of one of the most famous Muslim writers on mechanical devices (see Hill) had the full name Badīʿ al-Zamān Abu'l ʿIzz Ismaʿīl b. al-Razzāz al-Jazarī. Here the *laqab* "Badīʿal-Zamān" means "prodigy of the Age", certainly a title a scientist might wish to earn, and the *kunya* al-Jazarī signifies a person coming from al-Jazīra, the country between the upper reaches of the Tiger and Euphrates rivers.

Exercises

Note: These exercises are suggestions for library research and are not intended to be answered on the basis of the information in this chapter. This is also true of several of the exercises in the following chapters.

1. Use the *Dictionary of Scientific Biography* to write a brief account of the lives and works of any of the following: (1) The Bānū Mūsā, (2) al-Kindī, (3) Kamāl al-Dīn Fārisī or (4) Naṣīr al-Dīn al-Ṭūsī.
2. Write a short paper giving an account of the lives and works of some of the translators mentioned in this chapter, or of other important translators you have come across in your reading.
3. Write a short account of the major astronomical observatories in Islam.
4. Where were the major centers of political and military power in the Islamic world between the years 700 and 1400?
5. What are the main features of the calendar in use in Baghdad in the tenth century?
6. How did the man who called out the times for the five daily prayers know when it was time?
7. Write a short account of any of the following topics: (1) The book trade, (2) Education, (3) Libraries in Islam. (For the first and third consult Pedersen. For the second consult G. Maqdisi, *The Rise of Colleges*. Edinburgh: Edinburgh University Press, 1981.)
8. What have been the relative roles of anthropomorphic, zoomorphic and geometric art in Islam?

Bibliography

The following texts are basic references, which are useful for further reading on any of the topics in this book.

Gillespie, C. C. *et al.* (eds.). *Dictionary of Scientific Biography* (16 vols.). New York: Charles Scribner's Sons, 1972–80.

Sezgin, F. *Geschichte des arabischen Schrifttums* (Vol. 5 (Mathematics) and Vol. 6 (Astronomy)). Leiden: E. J. Brill, 1974 and 1978, resp.

This is, for the period to roughly 1050, the standard bio-bibliographical survey of Arabic literature.

Storey, C. A. *Persian Literature: A Bio-Bibliographical Survey*. Luzac and Co., 1927.

Wensink, H. *et al.* (eds.). *Encyclopedia of Islam*, 2nd ed. Leiden: E. J. Brill, 1960–present.

The second edition of this basic reference work is appearing in fascicles, but the first edition (pub. 1913–1934) is complete and is still valuable.

Some important general sources for further study of the history of mathematics in Islam are the following:

Kennedy, E. S. "The Exact Sciences in Iran under the Saljuks and Mongols". In: *Cambridge History of Iran*, Vol. 5. Cambridge, U.K.: Cambridge University Press, 1968, pp. 659–679.

Kennedy, E. S. *et al. Studies in the Islamic Exact Sciences*. Beirut: American University of Beirut Press, 1983.

This book contains a collection of papers on topics in the history of mathematics, astronomy and astrology, and geography in Islam by one of the most distinguished workers in the field, as well as by his students.

Kennedy, E. S. "The Arabic Heritage in the Exact Sciences", reprinted in Kennedy *et al. Studies in the Islamic Exact Sciences*. Beirut: American University of Beirut Press, 1983, pp. 30–47.

Youschkevitch, A. P. *Les Mathématiques Arabes* (*VIIIe-XVe Siècles*) (transl. by. M. Cazenave and K. Jaouiche). Paris: J. Vrin, 1976.

The following bear more specifically on the material in Chapter 1.

Berggren, J. L. "Nine Muslim Sages", *Hikmat* **1** (No. 9) (1979), and "Mathematics in Medieval Islam", *Hikmat* **2** (1985), 12–16, 20–23.

Both contain biographical sketches of important mathematicians of the Islamic world.

Berggren, J. L. "Al-Bīrūnī on Plane Maps of the Sphere", *Journal for the History of Arabic Science* **5** (1981), 47–96.

Haywood, J. A. and H. M. Nahmad, *A New Arabic Grammar*, 2nd ed. London: Lund Humphries, 1976.

Hitti, Phillip. *The Arabs: A Short History*, 5th ed. New York: St. Martin's Press, 1968.

This is an abridgement of his *History of the Arabs*.

Kennedy, E. S. "A Letter of Jamshīd al-Kāshī to His Father", *Orientalia* **29** (1960), 191–213.

This is reprinted in pp. 722–744 of Kennedy *et al. Studies in the Islamic Exact Sciences*. Beirut: American University of Beirut Press, 1983.

Meyerhoff, M. "On the Transmission of Greek and Indian Sciences to the Arabs", *Islamic Culture* 11 Jan. (1937), 17–37.

This version of the German original "Von Alexandrien Nach Baghdad" shows how scientific knowledge was transmitted to the Islamic world.

Pedersen, J. *The Arabic Book* (transl. by G. French). Princeton, NJ: Princeton University Press, 1984.

Rosenthal, F. *The Classical Heritage in Islam*. Berkeley and Los Angeles, CA.: University of California Press, 1975.)

This collection of texts and accompanying commentary illustrates the extent of medieval Islam's acquaintance with the classical world.

Toth, Imre, "Non-Euclidean Geometry Before Euclid", *Scientific American*, Nov. (1969), 87–95.

Toomer, G. J. "Lost Greek Mathematical Works in Arabic Translation", *The Mathematical Intelligencer* **6** (No. 2) (1984), 32–38.

Tritton, A. S., *Arabic* (Teach Yourself Books). London: Hodder and Stoughton, 1975.

Chapter 2

Islamic Arithmetic

§1. The Decimal System

Muslim mathematicians were the first people to write numbers the way we do, and, although we are the heirs of the Greeks in geometry, part of our legacy from the Muslim world is our arithmetic. This is true even if it was Hindu mathematicians in India, probably a few centuries before the rise of Islamic civilization, who began using a numeration system with these two characteristics:

1. The numbers from one to nine are represented by nine digits, all easily made by one or two strokes.
2. The right-most digit of a numeral counts the number of units, and a unit in any place is ten of that to its right. Thus the digit in the second place counts the number of tens, that in the third place the number of hundreds (which is ten tens), and so on. A special mark, the zero, is used to indicate that a given place is empty.

These two properties describe our present system of writing whole numbers, and we may summarize the above by saying the Hindus were the first people to use a cipherized, decimal, positional system, "Cipherized" means that the first nine numbers are represented by nine ciphers, or digits, instead of accumulating strokes as the Egyptians and Babylonians did, and "decimal" means that it is base 10. However, the Hindus did not extend this system to represent parts of the unit by decimal fractions, and, since it was the Muslims who first did so, they were the first people to represent numbers as we do. Quite properly, therefore, we call the system "Hindu–Arabic".

As to when the Hindus first began writing whole numbers according to this system, the available evidence shows that the system was not used by the great Indian astronomer Āryabhaṭa (born in A.D. 476), but it was in use by the time of his pupil, Bhaskara I, around the year A.D. 520. (See Van der Waerden and Folkerts for more details.)

News of the discovery spread, for, about 150 years later, Severus Sebokht, a bishop of the Nestorian Church (one of the several Christian faiths existing in the East at the time), wrote from his residence in Keneshra on the upper Euphrates river as follows:

> I will not say anything now of the science of the Hindus, who are not even Syrians, of their subtle discoveries in this science of astronomy, which are even more ingenious than those of the Greeks and Babylonians, and of the fluent method of their calculation, which surpasses words. I want to say only that it is done with nine signs. If those who believe that they have arrived at the limit of science because they speak Greek had known these things they would perhaps be convinced, even if a bit late, that there are others who know something, not only Greeks but also men of a different language.

It seems, then, that Christian scholars in the Middle East, writing only a few years after the great series of Arab conquests had begun, knew of Hindu numerals through their study of Hindu astronomy. The interest of Christian scholars in astronomy and calculation was, in the main, due to their need to be able to calculate the date of Easter, a problem that stimulated much of the Christian interest in the exact sciences during the early Middle Ages. It is not a trivial problem, because it requires the calculation of the date of the first new moon following the spring equinox. Even the great nineteenth-century mathematician and astronomer C. F. Gauss was not able to solve the problem completely, so it is no wonder that Severus Sebokht was delighted to find in Hindu sources a method of arithmetic that would make calculations easier.

We can perhaps explain the reference to the "nine signs" rather than the ten as follows: the zero (represented by a small circle) was not regarded as one of the digits of the system but simply a mark put in a place when it is empty, i.e. when no digit goes there. The idea that zero represents a number, just as any other digit does, is a modern notion, foreign to medieval thought.

With this evidence that the Hindu system of numeration had spread so far by the year A.D. 662, it may be surprising to learn that the earliest Arabic work we know of explaining the Hindu system is one written early in the ninth century whose title may be translated as *The Book of Addition and Subtraction According to the Hindu Calculation.* The author was Muḥammad ibn Mūsā al-Khwārizmī who, since he was born around the year A.D. 780, probably wrote his book after A.D. 800.

We mentioned in Chapter 1 that al-Khwārizmī, who was one of the earliest important Islamic scientists, came from Central Asia and was not an Arab. This was not unusual, for, by and large, in Islamic civilization it was not a man's place (or people) of origin, his native language, or (within limits) his religion that mattered, but his learning and his achievements in his chosen profession.

The question arises, however, where al-Khwārizmī learned of the Hindu arithmetic, given that his home was in a region far from where Bishop

Sebokht learned of Hindu numerals 150 years earlier. In the absence of printed books and modern methods of communication, the penetration of a discovery into a given region by no means implied its spread to adjacent regions. Thus al-Khwārizmī may have learned of Hindu numeration not in his native Khwārizm but in Baghdād, where, around 780, the visit of a delegation of scholars from Sind to the court of the Caliph al-Manṣūr led to the translation of Sanskrit astronomical works. Extant writings of al-Khwārizmī on astronomy show he was much influenced by Hindu methods, and it may be that it was from his study of Hindu astronomy that he learned of Hindu numerals.

Whatever the line of transmission to al-Khwārizmī was, his work helped spread Hindu numeration both in the Islamic world and in the Latin West. Although this work has not survived in the Arabic original (doubtless because it was superseded by superior treatises later on), we possess a Latin translation, made in the twelfth century A.D. From the introduction to this we learn that the work treated all the arithmetic operations and not only addition and subtraction as the title might suggest. Evidently al-Khwārizmī's usage is parallel to ours when we speak of a child who is studying arithmetic as "learning his sums".

§2. Kūshyār's Arithmetic

Survey of *The Arithmetic*

As we have said, al-Khwārizmī's book on arithmetic is no longer extant, and one of the earliest works on Hindu numeration whose Arabic text does exist was written by a man named Kūshyār b. Labbān, who was born in the region south of the Caspian Sea some 150 years after al-Khwārizmī wrote his book on arithmetic. Although Kūshyār was an accomplished astronomer, we know very little about his life, but despite this personal obscurity his works exerted some influence, and his treatise on arithmetic, whose title means *Principles of Hindu Reckoning*, became one of the main arithmetic textbooks in the Islamic world.

Kūshyār's concise treatise is a carefully-written introduction to arithmetic, divided into two main parts. The first contains, after a brief introduction, nine sections on decimal arithmetic, beginning with "On Understanding the Forms of the Nine Numerals". In it the nine numerals are given in a form standard in the east, namely:

١ ٢ ٣ ۴ ۵ ٦ ٧ ٨ ٩

and the place-value system is explained. Zero is introduced as the symbol to be placed in a position "where there is no number". The Arabic word for zero, "ṣifr", means "empty" and it is the source, via French and Spanish, of our word "cipher". It is even the source, via Italian, of our word "zero".

The following chapters are "On Addition", "On Subtraction"—with halving treated in a special subsection, "On Multiplication", "On the Results of Multiplication", "On Division", "On the results of Division", "On the Square Root", and "On Checking", through casting out nines.

The sixteen sections of the second part contain an explanation of the arithmetic of a base-60 positional system, but the book concludes with a section that tells how to find the cube root of a number in the decimal system. The base-60 system, which is now called a *sexagesimal* system, was important to astronomers because angles were measured, and trigonometric functions were tabulated, according to this system, and because its unified treatment of whole numbers and fractions made calculations so much simpler. We shall say more of this later.

As we follow Kūshyār's explanation of the decimal system it is well to bear in mind that he was explaining arithmetic to people who would be computing not with pen or paper but with a stick (or a finger) on a shallow tray covered with fine sand, which we shall refer to as a "dust board". Thus it is not convenient to have several rows of figures, for small boards are more convenient to carry around, but it is easy to erase (by smoothing out the sand), and we shall see how the algorithms for the five arithmetic operations (addition, subtraction, multiplication, division and the extraction of square roots) were designed with this feature of the dust board in mind.

In the text of his book Kūshyār writes out, in words, all the names of the numbers, and it is only when he is actually exhibiting what is written down on the dustboard that he uses the Hindu-Arabic ciphers. A reason for this may be that explanations were considered as text and therefore written out in words, like any other text. The examples of what was written on the dust board, however, may have been viewed as illustrations, much like a diagram in a geometrical argument, and they were there to show what the calculator would actually see on the dust board.

Addition

As Kūshyār explains this, the numbers to be added are written in two rows, one above the other, so that places of the same value are in the same column. He gives the example of adding 839 to 5625 and, unlike our method, begins his addition by adding from the highest place common to both numbers, in this case the hundreds' place, down. At each stage the answer obtained so far replaces part of the number on the top. Figure 2.1 illustrates his steps, beginning with $56 + 8 = 64$, and an arrow ($\rightarrow$) shows that one display on the right replaces, on the dust board, that on the left. Thus, at any time, there are only two numbers on the dust board, arranged in columns, and, in the end, the answer has replaced the number on the top. Unlike our method, the method Kūshyār explains obtains the left-most digit of the answer first.

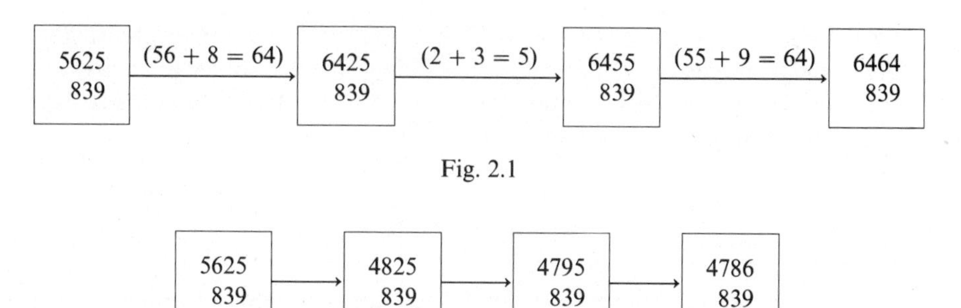

Fig. 2.1

Fig. 2.2

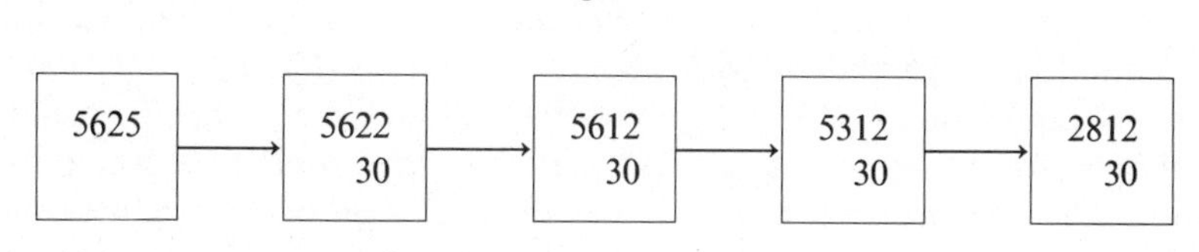

Fig. 2.3

Subtraction

Again Kūshyār explains the method by the same numbers, subtracting 839 from 5625, and again he works from left to right. He explains that since 8 cannot be subtracted from 6 it must be subtracted from 56 to produce 48, so the 56 of 5625 is erased and replaced by 48. Thus, working from place to place, Kūshyār obtains the answer, 4786 (Fig. 2.2). There is no "borrowing" in Kūshyār's procedure. He simply notices that, for example in the last step, since we cannot subtract 9 from 5 we must subtract it from 95. Just as with addition Kūshyār works from the higher places to the lower, and at each stage the partial answer appears as part of the number on top.

His treatment of halving, which he considers to be a variant of subtraction, sheds light on his treatment of fractions. He begins with 5625 (as usual), but this time he starts on the right (Fig. 2.3). He says to set down 5625 and then take half of five, which is two and a half. "Put two in the place of the five and put the $\frac{1}{2}$ under it, thirty."

He is using here, for fractions, the sexagesimal system, which goes back to the Babylonians and uses the principle of place value to represent fractions in terms of multiples of the subunits $1/60$, $1/60^2 = 1/3600$, etc. He explains the system more fully in the second part of the treatise, and here he contents himself with using his readers' familiarity both with the local monetary system in which a *dirham* contained 60 *fulūs*, and with degrees, in which 1 degree contains 60 minutes. Thus he tells his reader, in effect, "If you wish to think of 5625 as *dirhams* (degrees), then think of $\frac{5}{2}$ *dirhams* as 2 *dirhams* and 30 *fulūs* (2 degrees and 30 minutes)." The next two steps of his calculation, as shown in Fig. 2.3, are to halve the 2 in the 10's place and then the 6 in the

100's place, and now he must take half of the 5 in the 1000's place. Kūshyār values the 5 in terms of the preceding place, and so looks on it as 50 hundreds. Its half is thus 25 hundreds, and so in the last step he adds the 2500 to the half of 625 to obtain the answer shown in Fig. 2.3.

Multiplication

The algorithm for multiplication shows a thorough understanding of the rule for multiplying powers of 10, for to multiply 243 by 325 Kūshyār requires his reader to arrange the numerals so the 3 of 325 is directly above the 3 of 243 (Fig. 2.4). The total array occupies five columns, because hundreds multiplied by hundreds yields tens of thousands ($(N \cdot 100) \cdot (M \cdot 100) = N \cdot M \cdot 10{,}000$). Since $3 \cdot 2 = 6$, he places the 6 directly above the 2, i.e. in the ten thousands' place, and he remarks that had the product produced a two-digit number (e.g. had it been $4 \cdot 3 = 12$), the tens' digit of the product would be placed in the column to the left of the 2. This is illustrated at the next step where, since $3 \cdot 4 = 12$, he places the 2 of the 12 directly above the 4 and adds the 1 to the 6 to get 72. Finally, the top 3 is replaced by the $9 = 3 \cdot 3$, since he no longer needs to multiply by it.

Now we will be multiplying 243 by the upper 2, and since this counts "tens" and not "hundreds" we must, if we are to continue adding the answers to the top row in the columns above the bottom numerals, shift 243 one place to the right, since the powers of 10 represented by the answers will be one less. Thus we begin in the second row of Fig. 2.4, and as before, the last digit of the lower number (3) stands under the current multiplier (2). Then, since $2 \cdot 2 = 4$, we add the 4 to 72 to get 76, and the remaining steps of Row 2 will be clear to the reader who has followed those of Row 1. Again, a shift to the right automatically lines up the figures so that the answers are put

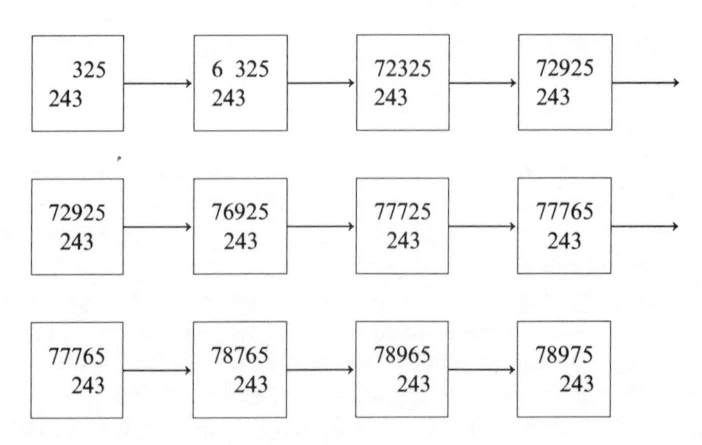

Fig. 2.4

in the correct place. We thus have Row 3 of Fig. 2.4, and the only thing to be careful of is that on the last multiplication of a given sequence (for example the final "$5 \cdot 3 = 15$") one does not add the final digit of the product to the last multiplier (in this case, 5), but, instead, uses it to replace the multiplier.

Division

This operation offers Kūshyār no more trouble than multiplication, as the division of 5625 by 243 shows. In Fig. 2.5 the first three steps show the process of subtracting $2 \cdot 243 = 486$ from 562, where the first digit of the quotient, 2, is obtained by an estimate and is written in the column above the last digit of the divisor. In this case, the "2" means 20, and the positioning automatically puts it in the right column. The last entry should be read $5625 - 20 \cdot 243 = 765$, and now Kūshyār moves the divisor over to the next column, so that the next digit of the quotient will be correctly aligned. The second row calculates $765 - 3 \cdot 243 = 36$, and so, calculating digit by digit, beginning with the one in the highest place, Kūshyār obtains the quotient (23) and remainder (36).

This answer, $23 + 36/243$, is correct but raises the further question, "How big is the fraction 36/243?". After all, an astronomer doing calculations with angles, or a judge dividing up a sum of money as an inheritance, needs the answer in a usable form. Thus a standard chapter in many arithmetics is one explaining how to express a fraction a/b in terms of some other subunits $1/c$, where c is a number appropriate to what is being measured. For example, if we were measuring lengths in feet and inches we would take $c = 12$, but Kūshyār proceeds to solve this problem for $c = 60$.

Of course, if $36/243 = n/60$, then $n = 36 \cdot 60/243$, and this division produces a quotient of 8 and a remainder of 216. So, if we think of the first remainder as being the *dirhams* left after the division of 5625 *dirhams* among 243 people, then each person's share would be 23 *dirhams* 8 *fulūs*, with 216 *fulūs* left over. Or, we could think of it as the division of an angle into 243 equal parts, so that each part would be $23°8'$ with $216'$ left over. This operation of multiplying a fraction, as 36/243, by 60 the Islamic authors called

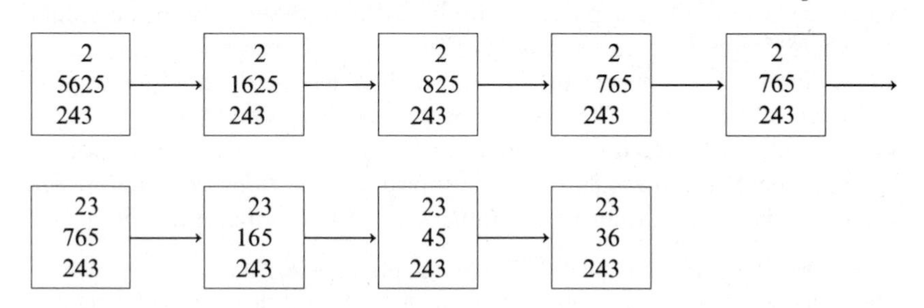

Fig. 2.5

"raising", and it was used to obtain base-60 expansions of the fractional parts in a division. It is the analogue of what we do to convert a fraction to percent.

§ 3. The Discovery of Decimal Fractions

Today we use not sexagesimal but decimal fractions to represent the fraction remaining after a division, and it now appears these were a contribution of the Islamic world. Evidence for this claim is contained in *The Book of Chapters on Hindu Arithmetic*, written in Damascus in the years A.D. 952–953 by Abu'l-Ḥasan al-Uqlīdisī. The name "al-Uqlīdisī" indicates that the author earned his living copying manuscripts of Euclid ("Uqlīdis" in Arabic), but beyond this we know nothing of the life of a man who seems to have been the first to use decimal fractions, complete with the decimal point, and therefore the first to write numbers as we do. Since al-Uqlīdisī specifically states in the preface to his book that he has taken great pains to include the best methods of all previous writers on the subject, it is hard to be sure that decimal fractions were his own discovery, but their complete absence in Indian sources makes it fairly certain that they were a discovery of Islamic scientists.

Al-Uqlīdisī is also proud of the fact that he has collected ways of performing on paper, with ink, algorithms usually performed by arithmeticians on the dust board, and in his *Book of Chapters* he gives the following reasons for abandoning the dust board in favor of pen and paper.

> Many a man hates to show the dust board in his hands when he needs to use this art of calculation (Hindu arithmetic) for fear of misunderstanding from those present who see it in his hands. It is unbecoming him since it is seen in the hands of the good-for-nothings earning their living by astrology in the streets.

It seems that the street astrologers could be recognized by their use of the dust board, and al-Uqlīdisī urges the use of pen and paper to escape being taken for a mendicant fortune teller.

Al-Uqlīdisī's text contains four parts, of which the first two deal with the elementary and advanced parts of Hindu arithmetic, but it is in the second part where decimal fractions first appear. This is in the section on doubling and halving numbers, where he introduces them as one of the three ways of halving an odd number. The first way is the one described by Kūshyār who, to halve 5625, considered it as degrees or *dirham*s and wrote the result as in Fig. 2.3, where the lower 30 could be interpreted as *fulūs* or minutes. The second way is one al-Uqlīdisī calls numerical, and describes as follows:

> ... halving one in any place is five (in the place) before it, and this necessitates that when we halve an odd number we make half of the unit five before it and we put over the units' place a mark by which we distinguish the place. So

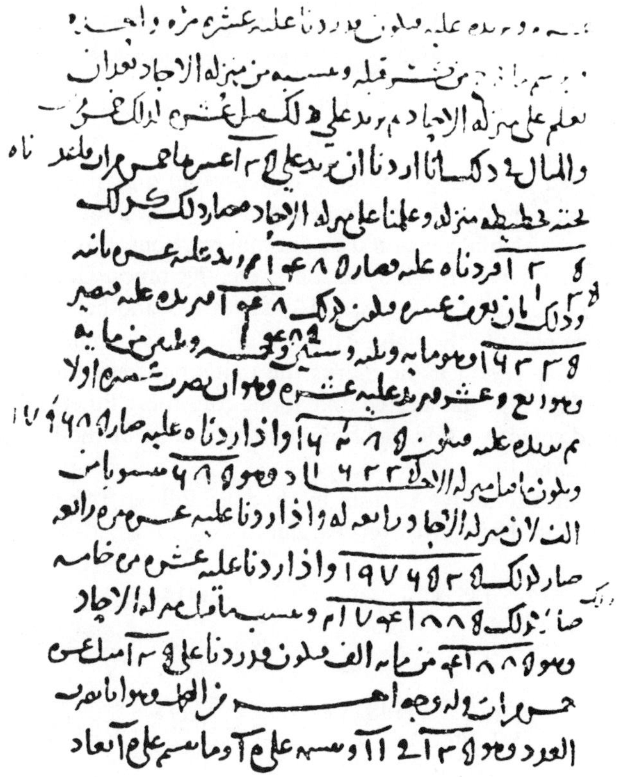

Fig. 2.6

the value of the unit's place is tens to that before it. Now the five may be halved just as whole numbers are halved, and the value of the units' place in the second halving becomes hundreds and this may continue indefinitely.

When al-Uqlīdisī writes of places in a numeral being "before" other places he is referring to the direction of Arabic writing, which is from right to left. Thus in 175 the 5 would be before the 7. As an example of what he has explained al-Uqlīdisī gives the results of halving 19 five times as 0́59375, where, he says, "the place of the units is hundred-thousands to what is in front of it". Figure 2.6 shows the Arabic text of al-Uqlīdisī's work and the use of the decimal point (in the form of the short vertical mark pointing out the unit's place). If the reader uses the forms of the numerals given earlier, he will have no trouble identifying the various numerals in that figure.

From a purely mathematical point of view it is especially satisfying to see decimal fractions (complete with decimal point) explained by reasoning by analogy from established procedures. Thus to halve 34 the usual procedure was $34 \to 32 \to 17$, where $(\frac{1}{2}) \cdot 30$ was reckoned as "$\frac{1}{2}$ of 3 is 1 and 5 in the place before it". What al-Uqlīdisī remarked was that the same principle

$$\begin{array}{lllllllllll}
13\overset{|}{5} & \longrightarrow & 14\overset{|}{8}5 & \longrightarrow & 16\overset{|}{3}35 & \longrightarrow & 17\overset{|}{9}685 & \longrightarrow & 19\overset{|}{7}6535 & \longrightarrow & 21\overset{|}{7}41885 \\
\;\;135 & & \;\;1485 & & \;\;16335 & & \;\;179685 & & \;\;1976535 & &
\end{array}$$

Fig. 2.7

could be applied to halving a number with an odd digit in the unit's place, and out of such a simple observation came a very useful mathematical tool.

A little later, al-Uqlīdisī again uses decimal fractions, this time to increase 135 by its tenth, then the result by its tenth, etc. five times. Thus he sets out as in Fig. 2.7 to calculate $135(1 + \frac{1}{10})^5$. He writes 135 and below it 135 again, but moved one place to the right. This will be $(\frac{1}{10}) \cdot 135$, so he adds it to 135. In the sum $135 + (\frac{1}{10}) \cdot 135$, he marks the unit's place with a short vertical line above it. When he has shifted and added four more times, the result will be the desired quantity. (He mentions that the value of the lowest place is hundred-thousandths.)

He gives an alternative to this method as follows (where we use the decimal point for al-Uqlīdisī's vertical line):

$$135 \cdot (1 + \tfrac{1}{10}) = \frac{(135 \cdot 11)}{10} = 148.5$$

and

$$148.5 \cdot (1 + \tfrac{1}{10}) = \frac{(148.5 \cdot 11)}{10} = 148 \cdot \left(\frac{11}{10}\right) + 0.5 \cdot \left(\frac{11}{10}\right)$$

$$= 162.8 + 0.55 = 163.35,$$

which shows that al-Uqlīdisī not only added decimal fractions but multiplied them by whole numbers as well, even though his method of multiplication unnecessarily splits the number into its whole and fractional parts.

Less than half a century later another Muslim author, Abū Manṣūr al-Baghdādī, used decimal fractions—also in a problem on computing tenths. He represented what al-Uqlīdisī would write as $1\overset{|}{7}\ 28$ by 08 02 17, but each pair written above the previous one, in strict analogy to Kūshyār's notation for sexagesimal fractions.

Al-Uqlīdisī's use of decimal fractions is something of an *ad hoc* device, unsystematized and unnamed. Two centuries later, however, one finds in the writings of al-Samaw'al, whose work we discuss in the chapter on algebra, the use of decimal fractions in the context of division and root extraction. In a treatise of 1172, al-Samaw'al introduces them carefully, as part of a general method of approximating numbers as closely as one likes. Thus al-Samaw'al uses decimal fractions within a theory rather than as an *ad hoc* device, although he still has no name for them and his notation is inferior to that of al-Uqlīdisī. The reader will find the details in Rashed.

It is in the early fifteenth century that decimal fractions receive both a name and a systematic exposition. By then Jamshīd al-Kāshī displays a thorough command of the arithmetic of decimal fractions, for example multiplying them just as we do today. It is also in the fifteenth century that a Byzantine arithmetic textbook describes as "Turkish", i.e. from the Islamic world, the method of representing $153\frac{1}{2}$ and $16\frac{1}{4}$ by 153|5 and 16|25 and their product by 2494|375. (See Hunger and Vogel.)

It was not until over a century later that the European writers began using decimal fractions. An able publicist for the idea was the Flemish engineer Simon Stevin, whose book *The Tenth* was published in 1585. However, his awkward notation was nowhere near so good as al-Uqlīdisī's, and it was left to the Scot, John Napier, to re-invent the decimal point and use decimal fractions in his table of logarithms, another invention of his.

§ 4. Muslim Sexagesimal Arithmetic

History of Sexagesimals

Although the student may think it strange that it took almost 500 years (from the tenth to the fifteenth centuries) for decimal fractions to develop, it must be remembered that Muslim scientists, from the ninth century onwards, already possessed a completely satisfactory place-value system to express both whole numbers and fractions. It was not decimal, however, but the *sexagesimal* system we have already referred to, in which the base is 60, and it arose out of the fusion of two ancient numeration systems.

The first of these is one used by the Babylonians around 2000 B.C in Mesopotamia. As we know it from the many surviving cuneiform texts, it was a positional system, in which the successive places of a numeral represent the successive powers (positive and negative) of the base, 60, so it treated whole numbers and fractions in a unified manner. However, the Babylonians did not use single ciphers for the fifty-nine digits from 1 to 59, but formed them by repeating the wedges for 1 (𒁹) and 10 (𒌋). Thus the Babylonians would represent the integers 3, 25, 133 and 3752 as

𒁹𒁹𒁹, 𒌋𒌋𒁹𒁹𒁹𒁹𒁹, 𒁹𒁹𒌋𒁹𒁹𒁹 $(2\cdot 60+13)$, 𒁹 𒁹𒁹𒌋𒌋𒌋𒁹𒁹𒁹 $(1\cdot 60^2+2\cdot 60+33)$.

In addition, they extended the system to include fractions. Thus since $\frac{1}{2}=\frac{30}{60}$ they would write $\frac{1}{2}$ as 30 also, 𒌋𒌋𒌋, and since

$$\frac{7}{360}=\frac{70}{3600}=\frac{60}{3600}+\frac{10}{3600}=\frac{1}{60}+\frac{10}{60^2}$$

it would be written as 𒁹𒌋.

There was always a possibility of misunderstanding in using this system, for there was no special mark to indicate the units (that is there was no sign like our decimal point) and no custom of writing final zeros in an integer, and therefore the magnitude of the number was determined only up to a factor of some power of 60. Thus, although 𒁹𒌋𒌋 could represent $1\frac{1}{3}$, it could also represent 80. One step towards clarification was taken late in the fourth century B.C., at the time when Babylon was ruled by the successors of Alexander. At that time, scribes in Babylon began to write numbers more frequently with a special symbol to indicate zeros within the numeral, so it was possible to write 71 in such a way as to distinguish it unambiguously from 3611 (71 being written as 𒁹𒌋𒁹 and 3611 as 𒁹𒑱𒌋𒁹).

These imperfections, however, are relatively minor and seemed not to have caused much difficulty for the Babylonians. Much more important is the existence, two millennia before our era, of a numeration system so well suited for complex calculations that Greek astronomers at some time during or after the second century B.C., adopted it for their calculations. Thus, the astronomer Ptolemy used it in the mid-second century A.D. in his Greek astronomical handbook *The Almagest*.

The Hellenistic Greeks' adoption of the system was, however, rather an instance of grafting than of transplanting; for, while they used it with a different notation to represent the fractional part of a number, they retained their own method of representing the integral part. This method is an example of the second ancient system we referred to earlier, in which 27 letters of an alphabet are used to represent the numbers 1, . . . , 9; 10, 20, . . . , 90; 100, 200, . . . , 900. In the case of the Greeks, they used 27 letters of an archaic form of the Greek alphabet, according to the scheme below:

Α	Β	Γ	Δ	Ε	Ϛ	Ζ	Η	Θ
1	2	3	4	5	6	7	8	9
Ι	Κ	Λ	Μ	Ν	Ξ	Ο	Π	Ϙ
10	20	30	40	50	60	70	80	90
Ρ	Σ	Τ	Υ	Φ	Χ	Ψ	Ω	Ϡ
100	200	300	400	500	600	700	800	900

This ancient alphabet stems from that of the Phoenecians, a Semitic people to whom we owe the inventions of the alphabet and of money. The alphabetic system of numeration seems to have been common to many of the peoples of the Mediterranean. Thus it was used not only by the Greeks and Arabs, but also by the Hebrews and others.

In this system the numbers up to 999 would be represented by a string of letters, so that, in the case of the Greeks 48 and 377 would be written MH and TOZ. We need not go into the special devices that were used to represent numbers larger than 999, for it is the fractions that interest us now. A Greek astronomer, knowing the Babylonian system, evidently saw the possibility of substituting letters of the alphabet for the groups of wedges the Babylonians used for digits. Thus $12\frac{1}{3}$ would be written *IB K*, to signify $(10 + 2) + \frac{20}{60}$.

The Greeks, however, adapted the Babylonian place-value system only for fractions, so they wrote *PMB IB* for $142\frac{1}{5}$ rather than the more consistent *B KB IB* (i.e. $2 \cdot 60 + 22 + \frac{12}{60}$). The only improvement the Greek system displayed was a slight cipherization for the digits, so that whereas the Babylonian would have to write 𒌋𒌋 for $\frac{1}{3}$, the Greek could simply write *K*.

The real transplant of the Babylonian system was done by Islamic mathematicians, in a system that was so widely used by astronomers that it simply became known as "the astronomers' arithmetic". In it, the 28 letters of the Arabic alphabet were used in an order quite different from their order in the alphabet as it was (and is) written. If we transcribe these letters according to the system in Haywood and Nahmad then the correspondence between letters of the Arabic alphabet and numerals is that shown in Fig. 2.8. (Although the system extends to 1000—for the 28th letter—there is no need for letters beyond the nūn (50) in the astronomers' arithmetic, since no digit can be greater than 59. The only fact we need to add is that, as with the Greeks, "zero" was represented by τ̈ or ȯ, which are two versions of the same cipher.)

Thus, if we represent Arabic letters by the corresponding Latin ones in Fig. 2.8, the numeral 84 would be written *a kd*, and *lb n* would represent $32\frac{50}{60}$. These two examples illustrate how the Muslims made a consistent adaptation of the Babylonian system to their own mode of writing, in the process of which they introduced a significant amount of cipherization. Of

ا	ب	ج	د	ه	و	ز	ح	ط
A	B	G	D	H	W	Z	Ḥ	Ṭ
1	2	3	4	5	6	7	8	9
ى	ك	ل	م	ن	س	ع	ف	ص
I	K	L	M	N	S	c	F	Ṣ
10	20	30	40	50	60	70	80	90
ق	ر	ش	ت	ث	خ	ذ	ض	ظ
Q	R	Sh	T	Th	Kh	Dh	Ḍ	Ẓ
100	200	300	400	500	600	700	800	900
				غ				
				Gh				
				1000				

Fig. 2.8

٢١ (21)
٠٥ (05)
٤٢ (42)

Fig. 2.9

course, there remained the ambiguity of the value of any given numeral. Although *b mh* could represent 165 ($= 2 \cdot 60 + 45$), it could equally well represent $2\frac{45}{60}$, and, in the absence of a sexagesimal point, some other device was necessary to eliminate this ambiguity.

There were two solutions to this problem. The one was to name each place, so that the nonnegative powers of 60 ($1, 60, 60^2, \ldots$) were called "degrees", "first elevates", "second elevates", ..., while the negative powers of 60 ($1/60, 1/60^2, 1/60^3, \ldots$) were called "minutes", "seconds", "thirds", ... The origin of the name "degrees" is in astronomy, where the term referred to the 360 equal parts into which the zodiac circle is divided. The term "minutes" is a translation of the Arabic *daqā'iq*, which means "fine", just as the English word "minute" does. The succeeding fine parts were, naturally, "the second, third, etc. fine parts". The other solution was to name the last place only, so that "*b mh* minutes" would make it clear that the value $2\frac{45}{60}$ was intended.

In the following survey of Muslim sexagesimal arithmetic we shall follow the second section of Kūshyār's *Principles of Hindu Reckoning*, and it is typical of the variety of approaches used by Muslim scientists that, although Kūshyār explains a consistent sexagesimal arithmetic, he does not use letters of the alphabet at all, but rather the form of the Hindu ciphers used in the Eastern caliphate. Thus, what some writers would express as *ka h mb*, Kūshyār writes as in Fig. 2.9, where the places of the numeral are written vertically in order to prevent confusion with the Hindu numeral 210,542. However, here as earlier he uses the ciphers only when he is actually showing the work. Elsewhere he writes out all the numerals longhand, and, to give some of the flavor of the work, we shall follow the same practice.

Sexagesimal Addition and Subtraction

To illustrate addition, Kūshyār gives the following example: "We wish to add twenty-five degrees, thirty-three minutes and twenty-four seconds to forty-eight degrees, thirty-five minutes and fifteen seconds." He sets these two numbers down in two columns, separated by an empty column, with degrees facing degrees, minutes facing minutes, and seconds facing seconds (Fig. 2.10). He then adds twenty-five to forty-eight, tens to tens and units to units, and then he adds thirty-three to thirty-five and twenty-four to fifteen. Whenever a sum exceeds sixty he subtracts sixty from it, enters the result, and adds one to the place above it. This is the reason for the upper "one"

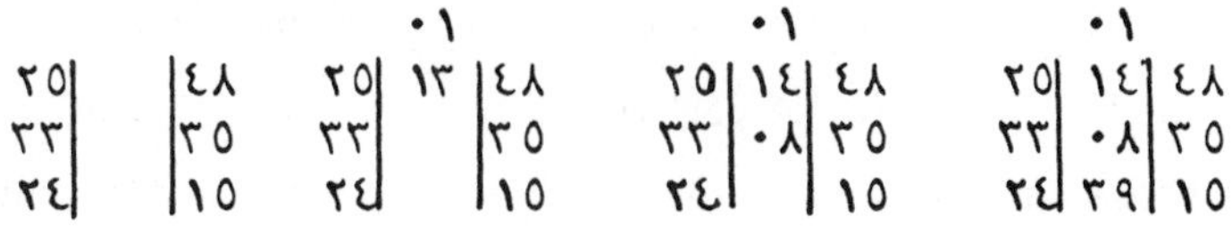

				٠١			٠١			٠١	
٢٥		٤٨	٢٥	١٣	٤٨	٢٥	١٤	٤٨	٢٥	١٤	٤٨
٢٣		٣٥	٢٣		٣٥	٢٣	٠٨	٣٥	٢٣	٠٨	٣٥
٢٤		١٥	٢٤		١٥	٢٤		١٥	٢٤	٢٩	١٥

Fig. 2.10

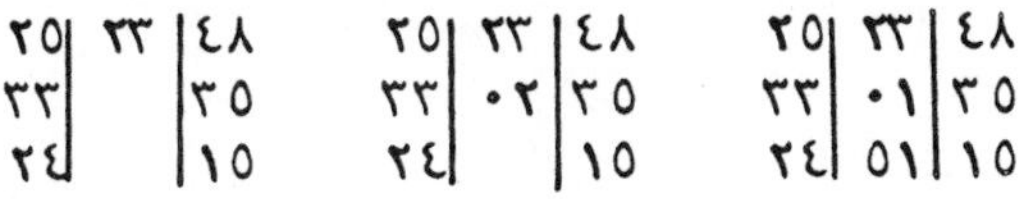

٢٥	٣٣	٤٨	٢٥	٣٣	٤٨	٢٥	٣٣	٤٨
٢٣		٣٥	٢٣	٠٢	٣٥	٢٣	٠١	٣٥
٢٤		١٥	٢٤		١٥	٢٤	٥١	١٥

Fig. 2.11

shown in the second figure. The dust board where Kūshyār imagined these calculations being carried out would show the successive parts of Fig. 2.10, with only the last set of figures showing at the end.

Subtraction, too, is straightforward, and it proceeds from the highest place downwards, with borrowing when necessary. Figure 2.11 shows the process of subtracting rather than adding in the above example, and it clearly offers no serious difficulties.

Sexagesimal Multiplication

Multiplication by Levelling

Multiplication and division were, however, another matter. Even such able mathematicians as al-Bīrūnī found it most convenient to convert the sexagesimal numerals to decimal form, perform the computations on the decimal forms by the rules of Hindu arithmetic, and then convert the answer back to sexagesimals, and the procedure was so common it was given a special name, "levelling". A contemporary of al-Bīrūnī, al-Nasawī, solves the problem of multiplying the two sexagesimal numbers $4°15'20''$ and $6°20'13''$ in the following way. First, he expresses both factors in terms of their lowest orders, Thus:

$$4°15' = (4 \cdot 60)' + 15' = 235', \quad \text{and} \quad 235'20'' = (235 \cdot 60)'' + 20'' = 14{,}120''.$$

Similarly, he calculates the other factor to be $22{,}813''$. Since the books that discuss this method explain how to calculate the products of various orders, al-Nasawī knows that the product of "seconds" by "seconds" will be on the order of "fourths" and, calculating in pure decimal numbers, he finds the product to be 322,119,560 fourths. Now it is necessary to perform the inverse operation of levelling, namely to "raise" this number to a sexagesimal expression, by dividing by 60. (We saw an example of this at the end of the treatment of division in the section on decimal arithmetic.) Thus, in this case,

$$322{,}119{,}560'''' = (5{,}368{,}659 \cdot 60 + 20)'''' = 5{,}368{,}659''' + 20''''.$$

Finally, proceeding as above, but now with the thirds, then the seconds, and finally the minutes, al-Nasawī obtains the answer $24°51'17''39'''20''''$.

The foregoing, inelegant procedure was widespread but by no means universal. Kūshyār, although he mentions it in his treatise as one method, explains how to multiply two sexagesimal numbers without any such conversion.

Multiplication Tables

At the beginning of his section on sexagesimal arithmetic, Kūshyār describes a sexagesimal multiplication table, which consists of 59 columns, each headed by one of the integers from 1 to 59, and each containing 60 rows. The column headed with the integer 39, for example, contains in its rows the multiples of 39, from $1 \cdot 39$ to $60 \cdot 39$. Although Kūshyār's book has no such table, examples of these tables have survived in other treatises (See King and Plate 2.1.) The right-most column of each page in such a table is headed "the number" and contains the alphabetic numerals, those from 1–30 usually appearing on the right-hand page and those from 31–60 on the left. The succeeding columns (going from right to left, as in Arabic writing) are headed by the alphabetic numerals between 1 and 60. (Of course only a certain number of them appear on each page, for reasons of space.) Each column gives, as we mentioned above, the first sixty multiples of the integer that stands at the top, and in general these multiples will need two sexagesimal digits to express them. For example, the product of 13 (*ij*) by 8 (*ḥ*) would be written with the two digit numeral we transliterate as *a md*. The first twelve rows of the three right-most columns in Plate 2.1, transliterated and then translated, are shown in Fig. 2.12. The eighth row below the heading for example means that $8 \cdot 13 = 1\ 44$ (104) and $8 \cdot 14 = 1\ 52$ (112). (We use the convention that $n\,m;\ r\,s$ means $n \times 60 + m + r/60 + s/60^2$. Another common convention separates the sexagesimal digits by commas, as $n, m; r, s$.

A remarkable example of a multiplication table was compiled, probably by a Turkish astronomer, around the year 1600 and gives the first 60 multiples of each two-place sexagesimal number from 00 01 to 59 59, so one can find directly from the table such products as $14\ 34 \cdot 19 = 36\ 46$. The table has 212,400 entries and fills a ninety-page booklet. Other astronomers, doubtless, found it more convenient to use the more limited tables and compute other products as needed by one of the algorithms we shall now describe.

The first of these differs only slightly from the method Kūshyār gives for the multiplication of two decimal numbers. In the sexagesimal case, the numbers are written vertically rather than horizontally, with an empty

Plate 2.1. Part of a sexagesimal multiplication table. The right-most column is headed "the number" and shows the alphabetic numerals from 1 to 12. The succeeding columns (from right to left, as in Arabic handwriting) are headed by the numerals 13, 14, ..., 18 and the entries underneath them give their multiples expressed as two-place sexagesimals. (See Fig. 2.12 for a transliteration and translation of the right-most three columns of this table.) (Photo courtesy of the Egyptian National Library.)

column left between them to contain the product, and Kūshyār's procedure for the product of $25°42'$ by $18°36'$ is shown in Fig. 2.13.

Methods of Sexagesimal Multiplication

The first two steps, Kūshyār specifies, are done with the aid of the multiplication table for 18, and since the 30 in the first step and the 12 in the second are of the same order they must be added in the product, so 30 is replaced by 42.

Since the product of minutes by minutes is seconds, the answer is $7^1 58°1'12''$ (that is, "7 first elevates, 58 degrees, 1 minute and 12 seconds").

ID	IJ	Al-ʿadad	14	13	The Number
ID–	IJ–	A	14 0	13 0	1
KH–	KW–	B	28 0	26 0	2
MB–	LT–	G	42 0	39 0	3
NW–	NB–	D	56 0	52 0	4
A I	A H	H	1 10	1 5	5
A KD	A H	W	1 24	1 18	6
A LH	A LA	Z	1 38	1 31	7
A TB*	A MD	Ḥ	1 52	1 44	8
B W	A NZ	Ṭ	2 6	1 57	9
B K	B I	I	2 20	2 10	10
B LD	B KJ	IA	2 34	2 23	11
B MH	B LW	IB	2 48	2 36	12

* An error for NB on the part of the scribe.

Fig. 2.12

```
18 |      | 25         18 |  7  | 25         18 |  7  | 25         18 |  7  |
   |      |     --->      | 30  |     --->      | 42  |     --->      | 42  |         --->
36 |      | 42         36 |     | 42         36 | 36  | 42         36 | 36  | 25
                                                                            |
                                                                            | 42

   |  7  |                  |  7  |                  |  7  |
18 | 42  |     --->      18 | 57  |     --->      18 | 58  |
36 | 36  | 25            36 | 36  | 25            36 | 01  | 25
         |                      |                      |     |
         | 42                   | 42                   | 12  | 42
```

Fig. 2.13

In the last two steps, 36 is one place lower than 18 so its products with 25 and 42 must be added to the column where the answer is taking shape, but one place lower than the corresponding products for 18.

A method, that was popular both in Islam and the West for decimal multiplication is illustrated in Fig. 2.14 with an example from Jamshīd al-Kāshī's *Calculators' Key*. The problem is to multiply 13 09 51 20 minutes by 38 40 15 24 thirds, and since the largest number of places is four a square is subdivided to form a lattice of 16 subsquares, each of which is divided as in Fig. 2.14 into two equal triangles. On the edges of the square that intersect at the top corner, the two factors are written so that the term of lowest order in one factor and that of highest order in the other factor are put at the top, each term of both factors being labelled by its orders. Then each square is filled in with the product of the two numbers on the outer edges opposite its sides, so that when this product has two ciphers the cipher of the highest order

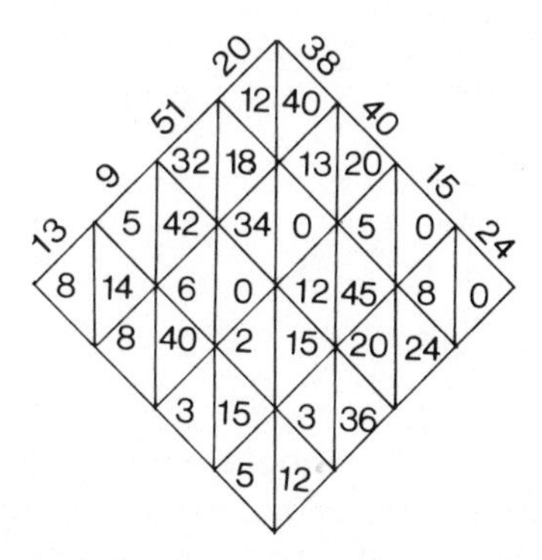

Fig. 2.14. The gelosia method of multiplication as described by al-Kāshī. Used for both decimal and sexagesimal systems it is shown here in the sexagesimal system. The answer is obtained by adding the eight columns within the square (base 60).

is put on the left of the square. For example, since $38 \cdot 13 = 8\ 14$ the 8 will be put in the left part of the left square and the 14 will be put in the right. When all 16 products are computed, the answer is obtained by adding up the ciphers in each of the eight vertical columns of the square, and the sums are written underneath. Since "minutes times thirds" is on the order of fourths, the lowest order of the product is fourths.

Although a certain amount of work is necessary to prepare the grid, it is then easy to fill in the lattice by means of a multiplication table, and the only computation involved is adding up the entries in the colums. Also, the squares can be filled in any convenient order, since the lattice-work keeps everything arranged.

The source from which we have taken this method, al-Kāshī's *The Calculators' Key*, gives no proof of the validity of the method; however, the proof is easy when one notices that what is put in the left-hand side of each square is precisely what would be carried and added to the next product in the method we are used to. The lattice does this carrying automatically, but what is carried is added to the product after all the multiplications have been done, rather than during the process as we are used to.

Sexagesimal Division

Finally, the method Kūshyār uses for division in the sexagesimal system parallels that used in the case of multiplication. Hence, to divide $49°36'$ by $12°25'$, Kūshyār arranges three columns and proceeds as in Fig. 2.15. Here

$$49 - 3 \cdot 12 = 13 \quad \text{and} \quad 13\ 36 - 3 \cdot 25 = 13\ 36 - 1\ 15 = 12\ 21,$$

so each digit of the divisor is multiplied by the digit of the quotient (obtained by trial, as we do), and the result is subtracted from that part of the dividend

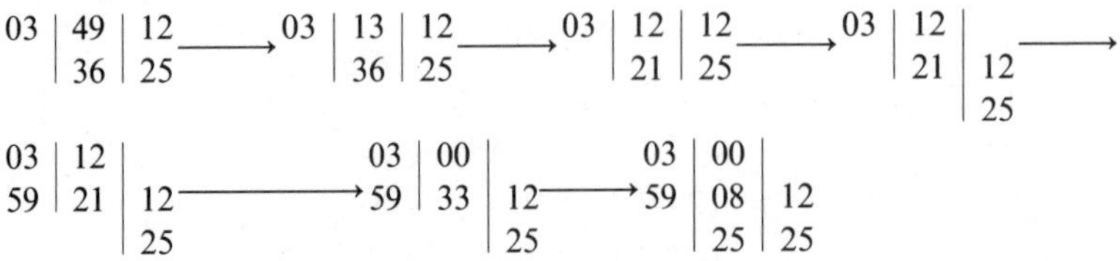

03 | 49 | 12 → 03 | 13 | 12 → 03 | 12 | 12 → 03 | 12 | →
36 | 25 36 | 25 21 | 25 21 | 12
25

03 | 12 | 03 | 00 | 03 | 00 |
59 | 21 | 12 → 59 | 33 | 12 → 59 | 08 | 12
25 25 25 | 25

Fig. 2.15

that is above it (or level with it) and to the right. Finally, the divisor is shifted down one place, and this is done so that the next digit of the quotient, when it is placed level with the highest entry of the divisor, will be of the correct order (in this case "minutes"). After the third step, the question now becomes "12°25′ times how many minutes produces something not exceeding 12°21′?", and the answer is "59 minutes". The last two steps of Fig. 2.15 show the final working-out. Again the general rule is that the product of a digit of the quotient by a digit of the divisor is subtracted from all of the dividend to the right and above (or level with) the digit of the quotient.

Thus Kūshyār gives the result as 3°59′ and says, "If we wish precision we copy the divisor one place lower". Hence the result could be continued to as many sexagesimal places as necessary. Also, Kūshyār remarks that he has attached to his book a table giving "the results of the division", that is, the order that results when a number of one order (say "first elevates") is divided by that of another (say "thirds"). Finally, Kūshyār concludes his chapter with a discussion of how to calculate square roots in the sexagesimal system, an operation of some importance to astronomers.

Thus, there was widespread in the Muslim world a consistent system of sexagesimal arithmetic that permitted a unified treatment of both whole numbers and fractions. This system was supported by special tables, and it provided an approach to all the operations of arithmetic which was every bit as satisfactory as that of the (initially) less-developed system of decimal fractions.

§5. Square Roots

Introduction

Instead of following Kūshyār's presentation of the extraction of square roots, we shall follow that of Jamshīd al-Kāshī in his book *The Calculators' Key*, which we have already referred to as the work he wrote in Samarqand two years before his death in 1429. It is a compendium of elementary mathematics, including arithmetic, algebra and the geometry of measurement, which contains a thorough treatment of decimal fractions, a table of binomial coefficients, and algorithms for extracting higher roots of numbers.

For example, we shall see later how he works out the fifth root of a number on the order of trillions, namely 44,240,899,506,917.

The following list of titles of the five main chapters of *The Calculators' Key* shows its differences from the work of Kūshyār: (1) On the arithmetic of whole numbers. (2) On the arithmetic of fractions (including decimal fractions). (3) On the arithmetic of astronomers (sexagesimal). (4) On the measurement of plane and solid figures. (5) On the solution of problems by algebra.

Obtaining Approximate Square Roots

We shall first see how al-Kāshī extracts the square root of 331,781. His method for the square root is no different from Kūshyār's, but, unlike Kūshyār, al-Kāshī was writing for people who would use pen and paper. (It was in Samarqand where the Arabs first learned of papermaking from Chinese prisoners of war near the end of the eighth century and, because of its abundant supply of fresh water, Samarqand remained a center of paper manufacturing for several centuries.) Thus, in the method as al-Kāshī explains it, none of the intermediate steps are erased. Al-Kāshī organizes his work by dividing the digits of the radicand, 331,781, into groups of two called "cycles", starting from the right. (Thus 331,781 is divided as 33 17 81.) As al-Kāshī explains it, since the numbers 1, 100, 10,000 ... have integer square roots (unlike 10, 1000, ...), the cycles are relevant, for the first (81) counts the number of units, the second (17) the number of hundreds, the third (33) the number of 10,000's, etc. He then draws a line across the top of the radicand and lines down the paper separating the cycles. At the beginning, therefore, his paper looks like Fig. 2.16(a).

To get the first of the three digits of his root he finds the largest digit n so that n^2 does not exceed 33. Since $5^2 = 25$ and $6^2 = 36$ he takes $n = 5$, which is written both above and some distance below the 33 (below the last 3) to obtain Fig. 2.16(b).

Now he subtracts 25 from 33 to obtain 8, which he writes below 33, and draws a line under 33 to show he is done with it. (On the dust board the 33

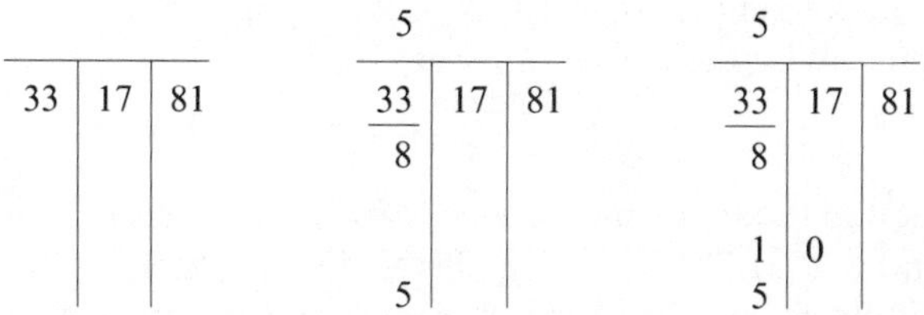

33	17	81

(a)

5		
33	17	81
8		
5		

(b)

5		
33	17	81
8		
1	0	
5		

(c)

Fig. 2.16

would be erased and the 8 would replace it.) Now he doubles the part of the root he has obtained, 5, and writes the result (10) above the bottom 5, but shifted one place to the right to obtain Fig. 2.16(c). (The dust board would only show the top 5, the middle 8 17 81, and the bottom 10 of Fig. 2.16(c).) At this stage al-Kāshī has a current answer (5) at the top and double the current answer (10) at the bottom.

What al-Kāshī next asks is to find the largest digit x so that $(100 + x) \cdot x \leq 817$. Experiment shows $x = 7$, and he writes 7 above the 7 of 17, and next to the 10 on the bottom, and then he performs the computation of $(100 + 7) \cdot 7 = 749$ and subtracts the result from 817 to get Fig. 2.17(a). He now begins the process again, doubling the last digit in 107 to get 114 and writing this above the 107, but shifted one place to the right, as shown in Fig. 2.17(b). Once again he has the current answer (57) at the top and double that (114) at the bottom, and as before the question is this: What is the largest digit x so that $(1140 + x) \cdot x \leq 6881$? A trial division of 688 by 114 suggests trying $x = 6$. This works, and after $1146 \cdot 6$ has been subtracted from 6881, the last digit of 1146 is doubled to make the 1146 into 1152 (= 1146 + 6) as in Fig. 2.17(c). (The dust board would show the top 576, the middle 5 and the bottom 1152.)

Thus al-Kāshī has obtained an approximate square root (576) as the current answer and double that (1152) at the bottom. Finally, he increases 1152 by 1 to get 1153 and divides it into the remainder, the middle 5, to obtain, as the approximate square root of 331,781, the number $576\frac{5}{1153}$ ($\doteq 576.00434$). Calculation shows the square of the latter number is 331,780.996, so al-Kāshī's result is quite close.

Justifying the Approximation

Two questions arise: (1) What is the justification for al-Kāshī's procedure for obtaining the integral part of the root; and (2) What is the justification for the fractional part? We will begin with the second question.

Justifying the Fractional Part

In fact, the numerator of the fractional part, 5, is equal to $331{,}781 - (576)^2$, and the denominator, 1153, is $577^2 - 576^2$. This is because 1153 is one more than "twice the current answer", i.e.

$$1153 = 1 + 2 \cdot 576 = (1 + 576)^2 - 576^2.$$

Thus the fractional part of the answer, $\frac{5}{1153}$, is just that obtained by *linear interpolation*, i.e. $(331{,}781 - 576^2)/(577^2 - 576^2)$, a technique that was ancient when Ptolemy used it in his *Almagest* in the first half of the second century A.D.

To understand this technique as a medieval astronomer might have justified it, imagine a table of square roots obtained by listing in one column the

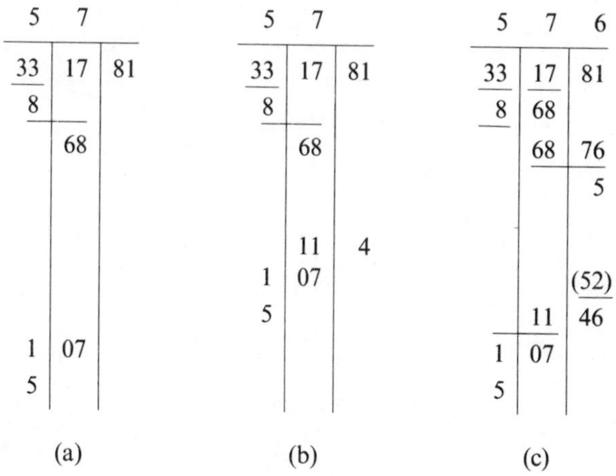

5	7	
33	17	81
8		
	68	
1	07	
5		

(a)

5	7	
33	17	81
8		
	68	
	11	4
1	07	
5		

(b)

5	7	6
33	17	81
8	68	
	68	76
		5
		(52)
	11	46
1	07	
5		

(c)

Fig. 2.17

N	$\sqrt{N}$
1	1
4	2
9	3
331,776	576
332,929	577
1,000,000	1,000

Fig. 2.18

successive squares from 1^2 to $1{,}000^2$, and, next to these in a second column, the first thousand whole numbers, as in Fig. 2.18.

Then, to find $\sqrt{6}$, the simplest procedure would be to observe that $4 < 6 < 9$ implies $2 < \sqrt{6} < 3$. Moreover, since $6 - 4 = 2$ and $9 - 4 = 5$, 6 is $\frac{2}{5}$ of the way between 9 and 4. Thus $\sqrt{6}$ is about $\frac{2}{5}$ of the way between $\sqrt{4} = 2$ and $\sqrt{9} = 3$, i.e. $\sqrt{6} = 2\frac{2}{5}$, approximately.

The reader will recognize that this reasoning is based on the assumption that $\sqrt{x}$ is proportional to x, which is the same as the assumption that, if we may express ourselves in modern language, the function $f(x) = \sqrt{x}$ is linear, i.e. its graph is a straight line. Although this is not true a glance at the graph of $f(x)$ in Fig. 2.19 reveals that it is nearly linear for $x > 1$ and over not too big an interval $[a, b]$. Thus, for example, the straight line joining the two points (16, 4) and (25, 5) is hardly distinguishable from the graph between these two points, and this is why the technique gives such a good approximation to the fractional part of $\sqrt{331{,}781}$. The table shows $576^2 = 331{,}776 < 331{,}781 < 332{,}929 = 577^2$ and, since $331{,}781 - 576^2 = 5$ while $577^2 - 576^2 = 1153$ we conclude that since $N = 331{,}781$ is $\frac{5}{1153}$ of the way between

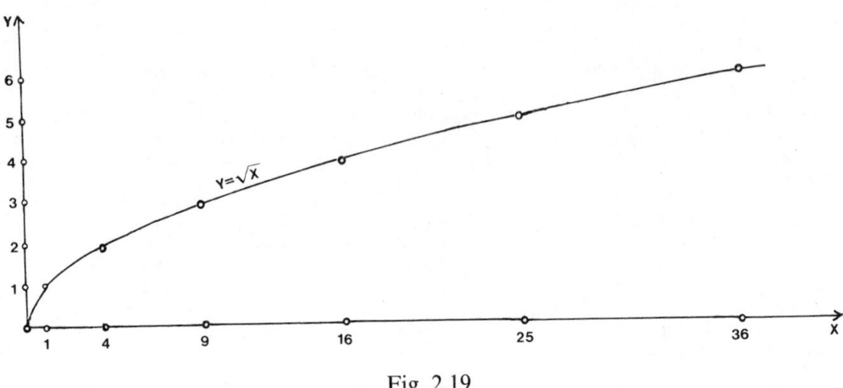

Fig. 2.19

576^2 and 577^2 its square root is about $\frac{5}{1153}$ of the way between 576 and 577. Thus $\sqrt{331{,}781} = 576\frac{5}{1153}$.

Linear interpolation was one of the standard ancient and medieval methods of what has been called "reading between the lines" (of tables), but we emphasize that the name "linear interpolation" reflects modern ideas, and those who discovered this method of approximation had no conception of a straight line as the graph of an equation. The ancient and medieval concept was simply that in a table of values pairing x to y one assumed that the change from y to y' was distributed equally over the units from x to x'.

Justifying the Integral Part

As for the extraction of the integral part of $\sqrt{N}$, al-Kāshī knows that if $N = abcdef$ then the largest integer r with the property that $r^2 \leq N$ has half as many digits as N, in this case 3. (By taking $a = 0$ if necessary we may assume that N has an even number of digits.) Therefore he divides N into what he calls cycles, as $N = ab\ cd\ ef$, and thereby considers $N = ab \cdot 100^2 + cd \cdot 10^2 + ef$.

Now, al-Kāshī's first step is to find the largest number A so that $A^2 \leq ab$. A will be a one-digit number, since ab has two-digits and 10^2 has three digits. Such an A will be the first digit of the root, as the reader may verify.

His next step is to calculate the difference

$$\Delta_1 = N - (A \cdot 100)^2 = (ab - A^2) \cdot 100^2 + cd \cdot 10^2 + ef$$

and then to find the next place, i.e. the largest B so that

$$\Delta_2 = N - (A \cdot 100 + B \cdot 10)^2 \geq 0.$$

He uses the basic identity $(X + Y)^2 = X^2 + (2X + Y)Y$ to expand

Δ_2 as $N - (A \cdot 100)^2 - (2A \cdot 100 + B \cdot 10)B \cdot 10 = \Delta_1 - (2A \cdot 10 + B)B \cdot 100$,

and the expression $2A \cdot 10 + B$ is the formal equivalent of al-Kāshī's instruction to double A, the previous digit of the root, and then put the digit (B) next to it. As al-Kāshī says, this next digit is chosen to be the largest so that the product $(2A \cdot 10 + B)B \cdot 100$ does not exceed the previous difference Δ_1. The multiplication of $2A$ by 1000 instead of by 10,000 is reflected in its shift one place to the right. Of course, al-Kāshī never mentions the powers of 10 since they are automatically taken into account by the positioning.

The procedure should by now be clear. When we have determined B to be as large as possible so that $(A \cdot 100 + B \cdot 10)^2 \leq N$ we choose C to be as large as possible so that $0 \leq N - (A \cdot 100 + B \cdot 10 + C)^2$, where, with $X = (A \cdot 100 + B \cdot 10)$ and $Y = C$, $(X + Y)^2$ is again expanded according to the rule $(X + Y)^2 = X^2 + (2X + Y)Y$. This identity, or its alternate form $(X + Y)^2 - X^2 = (2X + Y)Y$ is the basis for the algorithm for the extraction of the square root. Al-Kāshī's procedure also takes advantage of the fact that in evaluating $N - (X + Y)^2$ the part $N - X^2$ has been evaluated at the previous step.

§6. Al-Kāshī's Extraction of a Fifth Root

Introduction

We now follow the beginning of al-Kāshī's extraction of the fifth root of 44,240,899,506,197—a number on the order of trillions. The extraction of higher roots of numbers was, according to the testimony of ʿUmar Khayyām, an achievement of Muslim scholars, for he wrote in his *Algebra*,

> From the Indians one has methods for obtaining square and cube roots, methods which are based on knowledge of individual cases, namely the knowledge of the squares of the nine digits 1^2, 2^2, 3^2 (etc.) and their respective products, i.e. $2 \cdot 3$ etc. We have written a treatise on the proof of the validity of those methods and that they satisfy the conditions. In addition we have increased their types, namely in the form of the determination of the fourth, fifth, sixth roots up to any desired degree. No one preceded us in this and those proofs are purely arithmetic, founded on the arithmetic of *The Elements*".

ʿUmar was neither the first mathematician nor the last who believed falsely that he was the originator of a method. In this case we know that Abu l-Wafāʾ, who flourished over 100 years before ʿUmar, in the late tenth century, wrote a work entitled *On Obtaining Cube and Fourth Roots and Roots Composed of These Two*. Of course, ʿUmar may not have known of Abu l-Wafāʾ's treatise, or it may be that Abu l-Wafāʾ simply pointed out that $\sqrt[4]{N} = (\sqrt{\sqrt{N}}\,)$ and, since $\sqrt[3]{N}$ was already known from the Indians, roots such as the twelfth, for example, $\sqrt[12]{N} = \sqrt[4]{\sqrt[3]{N}}$ could be calculated by known methods. Thus Abu l-Wafāʾ's work may have been less innovative than that of ʿUmar.

Row of the result			
Row of the number	4424	08995	06197
Row of the square–square Row of the second of the number			
Row of the cube Row of the third of the number			
Row of the square Row of the fourth of the number			
Row of the root Row of the fifth of the number			

Fig. 2.20

Laying Out the Work

However that may be, neither ʿUmar's treatise nor that of Abu l-Wafā' is extant, so we shall study al-Kāshī's method from Book III of his *Calculators' Key*. He begins by instructing the reader to write the number across the top of the page and to divide the number into cycles, which are, this time, successive groups of five digits beginning from the right. This is because the powers of 10 with perfect fifth roots are 1, 10^5, 10^{10}, etc. Next, al-Kāshī puts, between the cycles, double lines and between the individual digits single lines, all running down the length of the page, and then he puts a line above the number, on which he will enter the digits of the root.

Next, he divides the space below the number into five broad bands by means of horizontal lines. The top band contains the number, and the words "Row of the number" are written on the edge of this band. The band below it is called "Row of the square square" (the fourth power), and it is also given the name "Row of the second of the number". When this process is finished, the sheet will look as in Fig. 2.20, and everything is ready. It seems not too far from the algorithmic spirit of this procedure to look on the cells in Fig. 2.20 as locations in a computer's memory, and in keeping with this Fig. 2.21 shows a flow-chart for the root extraction that the reader may find useful to get an overview of the process.

The Procedure for the First Two Digits

Al-Kāshī' now proceeds as follows. The largest integer, a, whose fifth power does not exceed 4424 is 5, so he puts 5 in "Row of Result" (above the first cycle) and at the bottom of "Row of Root". Next, he puts 5^2 at the

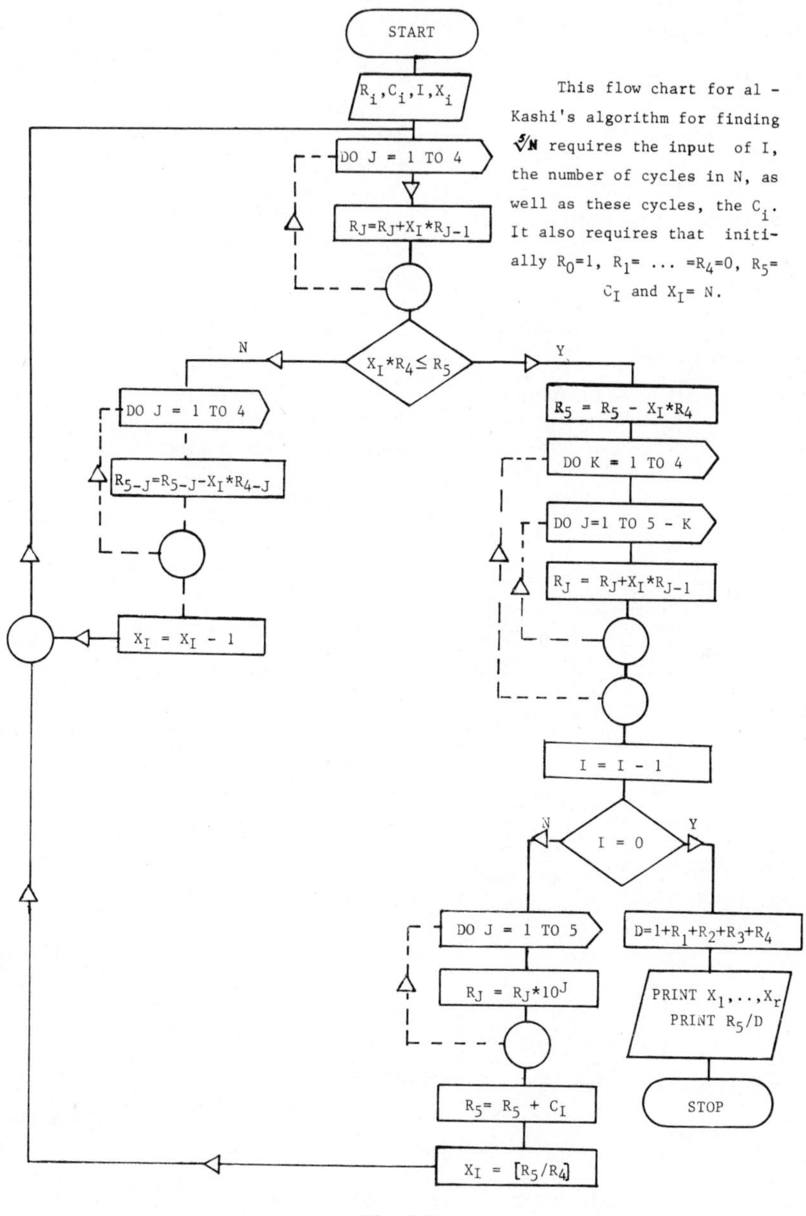

Fig. 2.21

bottom of "Row of Square", 5^3 at the bottom of "Row of Cube", and 5^4 at the bottom of the "Row of the Square Square". Finally, $4424 - 5^5 = 1299$ is placed in the "Row of Number" (This number, in virtue of its position, represents $1299 \cdot 10^{10}$.)

Next he begins the process called "once up to the row of the square

Row of the result	5		
Row of the number	4 4 2 4 3 1 2 5 1 2 9 9	0 8 9 9 5 = D	0 6 1 9 7
Row of the square–square Row of the second of the number	3 1 2 3 1 2 5 2 5 0 0 6 2 5	5 → Last step up to Row of Square Square	
Row of the cube Row of the third of the number	1 2 1 2 5 0 7 5 0 5 0 0 3 7 5 1 2 5	5 0 → Third step up to Row of Square Square	
Row of the square Row of the fourth of the number	2 5 0 1 0 0 1 5 0 7 5 7 5 5 0 2 5	2 5 0 → Second step up to Row of Square Square	
Row of the root Row of the fifth of the number	2 5 2 0 1 5 1 0 5	→ Start "Once up to Row of Square Square" 2 5	

Fig. 2.22

square", by adding the latest entry in "Row of Root" (5) to the most recently obtained digit of the root (5) and writing the sum (10) in "Row of Root" above 5. Now, he multiplies the sum by 5 and puts the product, $10 \cdot 5$, above 5^2 in "Row of Square" and then adds the two to get $75 = 5^2 + 50$. The sum he multiplies by 5 and puts the product $(75 \cdot 5)$ above 5^3 in "Row of Cube". Again, he adds these to get $500 = 5^3 + 75 \cdot 5$. Then he multiplies the sum by 5 and puts $500 \cdot 5$ above 5^4 in "Row of Square Square". Finally, he adds these to get $3125 = 5^4 + 500 \cdot 5$. (The lines within the bands mean, in the case of the bottom four bands, that all numbers below them would be erased on a dust board, and, in the case of the top band, the numbers *above* would be erased.)

Now, beginning with the 10 in "Row of Root", he repeats the above as far as "Row of Cube" ($10 + 5 = 15$, etc.), then, with 15, to "Row of Square" ($15 + 5 = 20$, etc.) and finally $20 + 5 = 25$ is put in "Row of Root".

Thus the numbers lying entirely in the first column are obtained. Now 3125 (in "Row of Square Square") is shifted one place right, 1250 (in "Row of Cube") two places, 250 (in "Row of Square") three places, and finally 25 (in "Row of Root") four places to the right, and he puts this "25" at the bottom of the next column (below the cycle 08995), as in Fig. 2.22.

At this point he seeks b, the largest single digit so that $f(b) \leq 129{,}908{,}995 = D$, where

$$f(b) = b(((b \cdot 25b + 250 \cdot 10^2)b + 1250 \cdot 10^3)b + 3125 \cdot 10^4),$$

where "$25b$" means $250 + b$.

It turns out that $f(4) = 146{,}665{,}024$ is too big, and since $f(3) = 105{,}695{,}493 < D$, al-Kāshī concludes that 3 is the desired value for b. (This method of evaluating a polynomial is standard in numerical analysis and is called Horner's method in many texts on the subject.)

Justification for the Procedure

The reader may easily verify with a pocket calculator that $530^5 < N$, while $540^5 > N$, and thus al-Kāshī has found the next digit of the fifth root. The question is: "How?", and the answer lies in the analogue of the identity that underlies extraction of square roots. If $C(n, k)$ denotes the binomial coefficient "n choose k", which counts the number of ways of choosing k objects from a set of n objects, then the binomial theorem applied to squares may be written

$$(A + B)^2 - A^2 = (C(2, 2)B + C(2, 1)A)B,$$

and, applied to higher powers, yields the identities:

$$(A + B)^3 - A^3 = ((C(3, 3)B + C(3, 2)A)B + C(3, 1)A^2)B,$$

$$(A + B)^4 - A^4 = (((C(4, 4)B + C(4, 3)A)B + C(4, 2)A^2)B + C(4, 1)A^3)B,$$

$$(A + B)^5 - A^5 = ((((C(5, 5)B + C(5, 4)A)B + C(5, 3)A^2)B + C(5, 2)A^3)B + C(5, 1)A^4)B.$$

The numbers $C(n, k)$ are arranged in a triangular array in Fig. 2.23. Notice that each row of this triangle begins and ends with a 1 and that a

							1							
						1		1						
					1		2		1					
				1		3		3		1				
			1		4		6		4		1			
		1		5		10		10		5		1		
	1		6		15		20		15		6		1	
1		7		21		35		35		21		7		1

Fig. 2.23

number greater than 1 in any row is just the sum of the two numbers to the right and left of it in the row above it. (Thus in the fourth row the "3" is the sum of 1 and 2 in the row above it.) If we begin numbering the rows with 0 and use the convention $C(0, 0) = 1$, then for all $0 \leq k \leq n$, $C(n, k)$ is the kth entry in Row n, and the rule for generating the triangle corresponds to the fundamental relationship

$$C(n, k) = C(n - 1, k) + C(n - 1, k - 1)$$

This triangle is called "Pascal's Triangle", after the French mathematician of the early seventeenth century, Blaise Pascal, whose *Traité du Triangle Arithmétique*, published in 1665, drew the attention of mathematicians to its properties. However, it might with more justice be called al-Karajī's triangle, for it was al-Karajī who around the year A.D. 1000 drew the attention of mathematicians in the Islamic world to the remarkable properties of the triangular array of numbers, a subject we shall return to in the chapter on algebra.

If we substitute the values of $C(5, k)$ into the expression for $(A + B)^5 - A^5$ we obtain the equality

$$(A + B)^5 - A^5 = ((((B + 5A)B + 10A^2)B + 10A^3)B + 5A^4)B.$$

In the present case, $A = 5 \cdot 10^2$ and $B = b \cdot 10$, and if we substitute these values for A and B the right-hand side of the expression now becomes

$$10^5((((b + \mathbf{25} \cdot 10)b + \mathbf{250} \cdot 10^2)b + \mathbf{1250} \cdot 10^3)b + \mathbf{3125} \cdot 10^4)b,$$

and the numbers in boldface are those appearing in the function f given earlier.

To see how al-Kāshī's technique generates the binomal coefficients, begin with a page divided into four horizontal bands, and, instead of writing the entries within a band one above the other, write them in a row, towards the right. Now, fill in the page as follows:

1. Put the first four powers of 1 up the left-most column, one in each band.
2. If any column has been filled in, start the next at the bottom by adding 1 to the entry to the left of it.
3. If any column has been filled in up to a given row, fill in the next row of

that column by adding 1 times the entry in the given row to that in the previous column of the next row.
4. Each column after the second contains one less row than the column to its left.

These rules will generate Fig. 2.24 in which the columns are just the diagonals descending to the right in Pascal's triangle, apart from the initial "1"s in these diagonals.

1	5			
1	4	10		
1	3	6	10	
1	2	3	4	5

Fig. 2.24

Of course, al-Kāshī wants not just the binomial coefficients $C(5, k)$, but the values $5C(5, 4) = 25$, $5^2C(5, 3) = 250$, $5^3C(5, 2) = 1250$ and $5^4C(5, 1) = 3125$. Thus, we construct a figure, on the model of Fig. 2.24, but this time:

1. Put ascending powers of 5 up the first column.
2. Add 5 instead of 1 as we move across the bottom row.
3. Whenever we move a number up to add it, first multiply it by 5.

Then we obtain Fig. 2.25 in which the last entries of the rows are the coefficients given in boldface in the expansion of f earlier.

625	3125			
125	500	1250		
25	75	150	250	
5	10	15	20	25

Fig. 2.25

One point remains, however. The numbers al-Kāshī must calculate with are not quite the above but $25 \cdot 10^6$, $250 \cdot 10^7$, $1250 \cdot 10^8$ and $3125 \cdot 10^9$. To represent these numbers on the array, al-Kāshī must move the 25 to the right four spaces, the 250 three spaces, the 1250 two, and the 3125 one. This is because, where they are, they are being treated as if they were multiplied by 10^{10}, while, in $f(b)$, 25 is multiplied only by 10^6. Since $10 - 6 = 4$, al-Kāshī must shift the 25 four places to the right to make it represent $25 \cdot 10^6$, etc. This is sufficient explanation of why, after al-Kāshī has found b according to the procedure outlined and has subtracted $f(b)$ from D, there remains $D' = N - (A + B)^5$.

The Remaining Procedure

Figure 2.26 shows the next part of the algorithm after 3 has been placed both in "Row of Result" and next to 25 in "Row of Root" (to form the number 253). The numbers in the parentheses on the right show how the algorithm calculates $f(b)$ in stages. Thus, 253 is multiplied by 3 to obtain 759, which is then put directly above the 25000 (the last two zeros not being

Row of the result	5	3	
Row of the number	1299 1056 243	08995 95493 = $f(3)$ 13502 = D'	06197
Row of the square–square Row of the second of the number	39 394 42 352 39 312	45240 52405 20574 31831 81831 } Fourth step in computing $f(3)$ 5	5
Row of the cube Row of the third of the number	 14 14 13 12	14887 88770 81912 06858 79581 27277 77277 } Third step in computing $f(3)$ 50	70
Row of the square Row of the fourth of the number		28 28090 786 27304 777 26527 768 25759 759 } Second step in computing $f(3)$ 250	090
Row of the root Row of the fifth of the number		265 262 9 6 253 } Begin computation of $f(3)$	265

Fig. 2.26

shown) and added to it to obtain 25,759. This is then multiplied by 3, written above the 1,250,000, and added to obtain 1,327,277. Finally, this is multiplied by 3 and the product added to the 31,250,000 in the row above it. This sum is finally multiplied by three and the product, which is $f(3)$, is subtracted from 129,908,995 in the "Row of the number". The difference, D', is $D - f(3)$.

Next, al-Kāshī begins the process "once up to the row of the square square" (with $253 + 3 = 256$, etc.), then to the row of the cube (with $256 + 3 = 259$), then to the row of the square ($259 + 3 = 262$). Finally, in the row of the number he puts $265 = 262 + 3$. The multiplications of course are all by 3 instead of by 5. The top numbers in the bands are then shifted so the numbers obtained will represent the constants in the polynomial:

$$g(c) = (((c \cdot 265c + 28{,}090 \cdot 10^2)c + 1{,}488{,}770 \cdot 10^3)c + 39{,}452{,}405 \cdot 10^4)c,$$

The next digit, c, must satisfy a condition entirely analogous to the one b satisfied, i.e. it must be the largest single digit so that $g(c)$ does not exceed $24{,}213{,}502 \cdot 10^5$. Al-Kāshī finds $c = 6$.

The bracketed lines in Fig. 2.27 denote the computation of the terms of $g(6)$, and D'' the final difference. Finally, al-Kāshī performs the procedure of going up to the "Row of Square Square", etc. The reader should now be able to follow without difficulty the steps as shown in Fig. 2.27.

The Fractional Part of the Root

At this point al-Kāshī has finished the calculation of the integer part of the fifth root of the given 14-place number. He had perfect control of decimal fractions and there is no doubt he knew that he could now shift again and continue the procedure to extract successive decimal places of the fifth root. Also, al-Khwārizmī, in a part of his treatise on arithmetic reported by the Latin writer John of Seville (fl. ca. 1140), gives an example of calculating $\sqrt{2}$ by calculating:

$$\sqrt{2} = \frac{\sqrt{2{,}000{,}000}}{1000} \doteq \frac{1414}{1000}$$

a procedure which al-Kāshī also recommends. So, even without decimal fractions, one can obtain any desired degree of accuracy.

What al-Kāshī does here, however, is to add up the top numbers in each of the four bottom rows and increase the sum by one, i.e. he forms

$$\begin{array}{r} 412694958080 \\ 1539906560 \\ 2872960 \\ 2680 \\ + \qquad\qquad 1 \\ \hline 414237740281 \end{array}$$

Row of the result	5	3	6	
Row of the number	2 4 2 {2 4 2	1 3 5 0 2 1 3 5 0 2	0 6 1 9 7 0 6 1 7 6 2 1	$= D''$
Row of the square–square	4 1	2 6 9 4 9	5 8 0 8 0	→ Last step to square square
Row of the second of the number	{4 0	9 1 3 6 5 3 5 5 8 3	9 0 3 8 4 6 7 6 9 6	→ Sum of two rows below
	3 9	9 0 3 4 3 4 5 2 4 0	1 7 6 9 6 ← 5	×6
Row of the cube		1 5 3 9 9	0 6 5 6 0	→ Last step to cube
Row of the third of the number		1 7 1 1 5 2 2 7	4 1 4 9 6 6 5 0 6 4	→ Third step to square square
		1 7 0 {1 5 0 5 7	4 5 4 4 8 1 9 6 1 6	→ Sum of two rows below
		1 6 9 1 4 8 8 7	4 9 6 1 6 ← 7 0	×6
Row of the square		2 8	7 2 9 6 0	
Row of the fourth of the number		2 8	1 6 0 4 4 5 6 9 1 6	→ Second step to row of cube
		2 8	1 6 0 0 8 4 0 9 0 8	→ Second step to square square
		{2 8	1 5 9 7 2 2 4 9 3 6	→ Sum of two rows below
		2 8	1 5 9 3 6 ← 0 9 0	
Row of the root			2 6 8 0	×6
Row of the fifth of the number			2 6 7 4	
			2 6 6 8	→ First step to row of cube
			2 6 6 2	→ First step to row of square square
			{2 6 5 6	→ Begin computation of $g(c)$

Fig. 2.27

and states that the fifth root of the given number is $536 + 21/414{,}237{,}740{,}281$.

Al-Kāshī's rule for finding the fractional part is based on the approximation

$$(n^k + r)^{1/k} \doteq n + \frac{r}{(n+1)^k - n^k}$$

where he explicitly calculates

$$(n + 1)^k - n^k = C(n, 1)n^{k-1} + C(n, 2)n^{k-2} + \cdots + 1.$$

This is, of course, just the method of linear interpolation we discussed earlier, for $n^k + r$ is r units of the total $(n + 1)^k - n^k$ units between two successive kth powers, so linear interpolation would place its kth root, $(n^k + r)^{1/k}$, the fraction $r/[(n + 1)^k - n^k]$ of the way between the two kth roots n and $n + 1$.

Figure 2.28 reproduces a page from the printed version of al-Kāshī's *The Calculators' Key*, which shows the entire calculation we have just explained. The reader will benefit from identifying the numerals and following the procedure through the first "Once up to the row of square square".

§7. The Islamic Dimension: Problems of Inheritance

Al-Khwārizmī devotes the first half of his book on algebra to solutions of the various types of equations and demonstrations of the validity of his methods, but the latter half contains examples of how the sciences of arithmetic and algebra could be applied to the problems posed by the requirements of the Muslim laws of inheritance.

When a person dies who leaves no legacy to a stranger the calculation of the legal shares of the natural heirs could be solved by the arithmetic of fractions. The calculation of these shares was known as *ʿilm al-farāʾiḍ*, and two examples from al-Khwārizmī's work illustrate the applications of arithmetic here.

The First Problem of Inheritance

This problem is a simple one, namely,

Example 1. "A woman dies, leaving her husband, a son and three daughters", and the object is to calculate the fraction of her estate that each heir will receive.

The law in this case is that the husband receives $\frac{1}{4}$ of the estate and that a son receives twice as much as a daughter. (It should be said, however, that from the woman's point of view Islamic inheritance law was a considerable improvement over what the pre-Islamic requirements in the Arabian peninsula had been.)

Al-Khwārizmī then divides the remainder of the estate after the husband's share has been deducted, namely $\frac{3}{4}$, into five parts, two for the son and three for the daughters. Since the least common multiple of five and four is twenty,

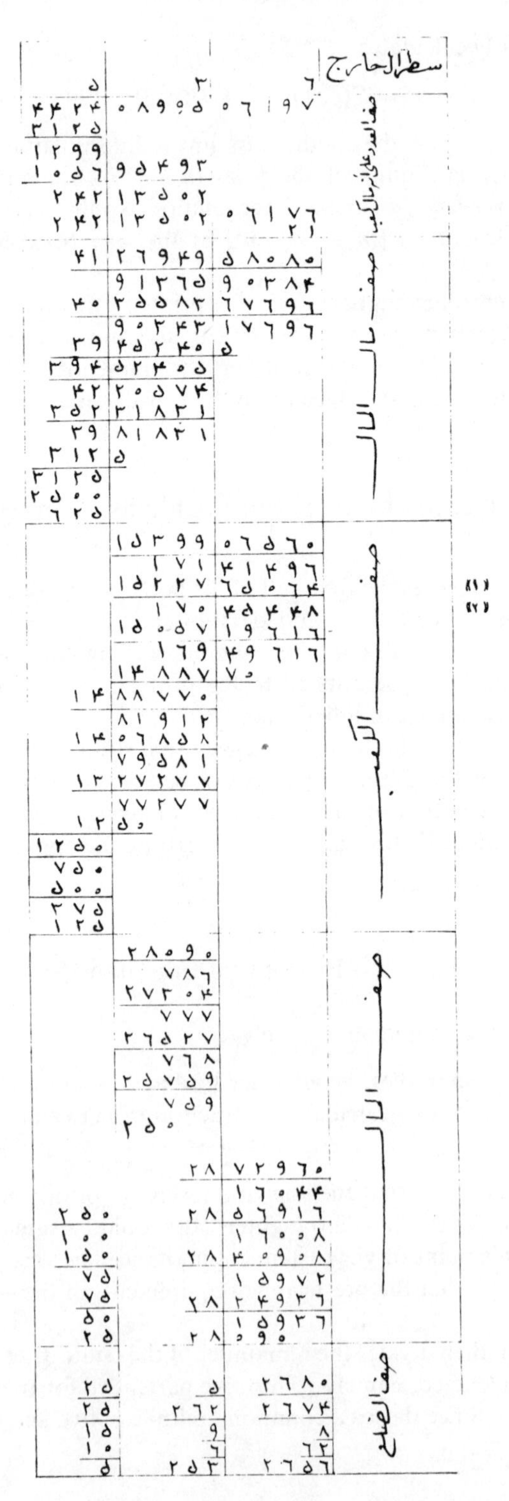

Fig. 2.28

the estate should be divided into twenty equal parts. Of these, the husband gets five, the son six, and each daughter three.

The Second Problem of Inheritance

This problem is a little more complicated and illustrates how unit fractions were employed to describe more complex fractions.

Example 2. A woman dies, leaving her husband, son and three daughters, but she also bequeaths to a stranger $\frac{1}{8} + \frac{1}{7}$ of her estate. Calculate the shares of each.

The law on legacies is that a legacy cannot exceed $\frac{1}{3}$ of the estate unless the natural heirs agree to it. (Here complications could enter because of the provision that if some agree and some do not agree those who do agree must pay, pro-rated, their share of the excess of the legacy over the third.) In the present case, however, since $\frac{1}{8} + \frac{1}{7} \leq \frac{1}{3}$ no complications enter, and the second provision on legacies, namely that a legacy must be paid before the other shares are calculated, now takes effect.

As in the above problem, the common denominator of the legal shares of her relatives is 20. Also, the fraction of the estate remaining after the stranger's legacy ($\frac{1}{8} + \frac{1}{7} = \frac{15}{56}$) has been paid is $\frac{41}{56}$. Then the ratio of the stranger's share to the total shares of the family is $(\frac{15}{56}):(\frac{41}{56}) = 15:41$. Thus, of the whole estate, the stranger will receive 15 parts to the 41 parts the natural heirs will receive. Multiplying both numbers by 20 to facilitate the computation of the shares of the heirs, we find that of a total of $20 \cdot (15 + 41) = 20 \cdot 56 = 1120$ parts the stranger receives $20 \cdot 15 = 300$ and the heirs jointly receive $20 \cdot 41 = 820$. Of these parts, the husband receives $\frac{1}{4}$, namely 205, the son $\frac{6}{20}$, namely 246, and each of the daughters gets 123.

On the Calculation of *Zakāt*

Another example of the use of arithmetic in the Islamic faith is in the calculation of *zakāt*, the community's share of private wealth. This is payable each year, at a certain rate, and the following problem, taken from *The Supplement of Arithmetic* of the eleventh-century mathematician, Abū Manṣūr al-Baghdādī, follows the gradual diminution of a sum of money as the *zakāt* is paid for three years. Its treatment of the fractional parts of a *dirhām* reminds one of Kūshyār's treatment of fractions, and in presenting it we paraphrase slightly, following the translation in Saidan (to appear) "We want to pay the *zakāt* on 7586 *dirhāms*, the amount that Muḥammad ibn Mūsā al-Khwārizmī mentioned in his work". (The *dirhām* was divided into sixty *fulūs*, the plural of *fils* (see Plate 2.2).)

The rate of *zakāt* is 1 *dirhām* in 40, but al-Baghdādī does not divide 7586

Plate 2.2. Obverse and reverse of two coins from the medieval Islamic world. The one on the right is a *fils* of Damascus, minted in 87 A.H. (Anno Hijra). The obverse of the coin on the right names the Caliph al-Walīd (of the Umayyad Dynasty) and gives the *shahāda* (Muslim confession of faith: "There is no god but Allah and Muḥammad is the Messenger of Allāh.") The one on the left is a *dirham* of Medīnat al-Salām (Baghdad) issued in 334 A.H. and names, on the two faces of the coin, the Būyid rulers Muʿizz al-Dawla and ʿImād al-Dawla as well as the Caliph al-Muṭīʿ. (Photo courtesy of the American Numismatic Society, New York.)

7586	7585	7583	7571	7396	7031
	34	34	14	14	6
					8
	(a)	(b)	(c)	(d)	14
					(e)

Fig. 2.29

by 40 according to the algorithm Kūshyār describes. Rather he calculates the total due on 7586 *dirhāms*, place-by-place, as follows:

From the first place we remove 1, which we make 40, and then remove 6 from the 40. This 6 is the *zakāt* due on 6 *dirhāms* and it is 6 parts of (the 40 into which we have divided) a *dirhām*. Thus, of the 40 there remains 34 parts. This we put under the five that has remained in the units place, as in Fig. 2.29(a).

We must now calculate $\frac{1}{40}$ of the 80 that arises from the 10's place, to obtain 2, which we subtract from the five in the unit's place. This leaves what is shown in Fig. 2.29(b).

In the 100's place there is 500, on which the *zakāt* due is $12\frac{1}{2}$. Of the 40 parts into which we have divided the *dirhām*, $\frac{1}{2}$ is 20, so when we subtract this from 34 there remain 14 parts. Also 12 from 83 leaves 71, so there now remain the figures shown in Fig. 2.29(c).

Finally, $\frac{1}{40}$ of the 7000 we obtain from the 1000's place is 175, and when we subtract this from 571 there remains 396, so the answer is that shown in Fig. 2.29(d).

Al-Baghdādī follows this for two more years, after which there remain the number of *dirhāms* shown in Fig. 2.29(e), where, e.g. the 14 means $14/(40)^3$ *dirhāms*. (The tax collector is going to get every last *fils* due!)

Of course, *dirhāms* are divided into 60 *fulūs*, not 40, and so, to calculate the *zakāt*, the base-40 fractions, which were convenient to use in the intial stage, must now be converted into sexagesimal fractions, and here al-Baghdādī points out a slip on the part of al-Khwārizmī, his source for the problem. Evidently, al-Khwārizmī said that if each of the fractional parts (i.e. 6, 8 and 14) is increased by $\frac{1}{2}$ then they become sexagesimal parts, i.e. minutes, seconds and thirds. This is of course true for the 6, because

$$\frac{6}{40} = 6 \cdot \frac{\frac{3}{2}}{40 \cdot \frac{3}{2}} = \frac{9}{60}$$

but it is false for the following parts, and al-Baghdādī gives the correct rule.

Without question, *'ilm al-farā'iḍ* is an important subject for Muslims, but in estimating the place of mathematics in that discipline a cautionary note, written by the great fourteenth-century Muslim historian from Tunis, Ibn Khaldūn, is worth recording.

> Religious scholars in the Muslim cities have paid much attention to it. Some authors are inclined to exaggerate the mathematical side of the discipline and to pose problems requiring for their solution various branches of arithmetic, such as algebra, the use of roots, and similar things. It is of no practical use in inheritance matters because it deals with unusual and rare cases. (Transl. in Rosenthal.)

Exercises

1. Use Kūshyār's method to add and subtract 12,431 and 987, showing your steps as in the text.
2. Try to develop an algorithm for halving a number that starts with the highest place in a number. Why do you think the Muslim calculators worked from the lowest place?
3. Use an operation modelled on raising to obtain a *decimal* expansion of $\frac{243}{7}$.
4. Use Kūshyār's method to multiply 46 and 243.
5. Use Kūshyār's method of division to divide 243 by 7, and then use the method of raising to find a 3-place sexagesimal approximation to $\frac{5}{7}$.
6. Adapt the method of raising to find a 3-place decimal approximation to $\frac{5}{7}$.
7. Devise a method for converting decimal integers to sexagesimal integers. Do the same for fractions. Now do the same, but going from sexagesimal to decimal.

8. List some possible values for *KE MB H*, including some fractional ones.
9. Add, subtract and multiply the two sexagesimal numbers 36,24 and 15,45. Divide 2,6,15,0 by 8,20.
10. Use the lattice method to multiply 2468 by 9753.
11. With A and N as in the section on square roots show that $((A + 1) \cdot 100)^2 > N$, while $(A \cdot 100)^2 < N$. Conclude that A is the first digit of the root.
12. Use al-Kāshī's method, including linear interpolation, to find $\sqrt{20{,}000}$.
13. If a man dies, leaving no children, then his mother receives $\frac{1}{6}$ and his widow $\frac{1}{4}$ of the estate. If he has any brothers or sisters, a brother's share is twice that of a sister. Find the fractions of the estate due if a man dies, leaving no children but a wife, a mother, a brother, two sisters and a legacy of $\frac{1}{9}$ of the estate to a stranger.
14. Give a rule for converting the remaining base-40 parts in the example from al-Baghdādī to sexagesimal parts. Generalize this rule to one for converting fractions from base n to base m.
15. Show that for any single digit b

$$f(b) = (5 \cdot 10^2 + b \cdot 10)^5 - (5 \cdot 10^2)^5$$

and conclude that b is the desired second digit of the fifth root, where f is the function in our discussion of al-Kāshī's extraction of the fifth root.

16. Al-Kāshī's method of evaluating $f(b)$ suggests evaluating an arbitrary polynomial

$$g(x) = a_n x^n + a_{n-1} x^{n-1} + \cdots + a_1 x + a_0$$

as

$$g(x) = (\ldots((a_n x + a_{n-1})x + a_{n-2})x + \cdots + a_1)x + a_0,$$

where the initial dots denote parentheses and those in the middle denote intermediate terms.

   (a) Evaluate $g(2)$, where $g(x) = 5x^3 - 3x^2 + 7x + 6$ by this method.
   (b) If addition and multiplication are each counted as one operation how many operations are necessary to evaluate $g(x)$ by this formula? How many are necessary according to the usual method? Which would use less computer time?

17. Show that the sum 412,694,958,080 + $\cdots$ + 1 calculated in al-Kāshī's extraction of the fifth root is equal to $537^5 - 536^5$.
18. Use al-Baghdādī's method and format (as in Fig. 2.29) to supply the details of the computation of the *zakāt* for years two and three.

Bibliography

Dakhel, Abdul-Kader. *Al-Kāshī on Root Extraction*. Sources and Studies in the History of the Exact Sciences, Vol. 2. W. A. Hijab and E. S. Kennedy, (eds.). Beirut: American University of Beirut, 1960.

Hunger, H. and K. Vogel. *Ein Byzantinisches Rechenbuch des 15. Jahrhunderts.* Wien: Hermann Boelhaus Nachf., 1963.

Ibn Labbān, Kūshyār. *Principles of Hindu Reckoning* (transl. and comm. M. Levey and M. Petruck). Madison and Milwaukee, WI: University of Wisconsin Press, 1965.

Al-Kāshī. *Miftah al-Hisab.* (edition, notes and translation by Nabulsi Nader). Damascus: University of Damascus Press, 1977.

This is the work of al-Kashī whose title we translate *The Calculators' Key.* Our Fig. 2.28 is taken from this book, courtesy of the publishers.

King, D. A. "On Medieval Multiplication Tables", *Historia Mathematica* **1** (1974), 317–323 and "Supplementary Notes on Medieval Islamic Multiplication Tables", *Historia Mathematica* **6** (1979), 405–417.

Rashed, R. "L'Extraction de la Racine n^{ieme} et l'invention des Fractions Décimales. XIe–XIIe siècles", *Archive for History of Exact Sciences* **18** (No. 3)(1978), 191–243.

Saidan, A. S. "Arithmetic of Abu l-Wafā'", *Isis* **65** (1974), 367–375.

This contains a summary of an important work on finger arithmetic among the Islamic peoples.

Saidan, A. S. "The Takmila fi'l-Ḥisāb by al-Baghdādī". In: *Festschrift: A Volume of Studies of the History of Science in the Near East, Dedicated to E. S. Kennedy.* (D. A. King and G. A. Saliba, eds.). New York: New York Academy of Sciences (to appear).

Al-Uqlīdisī, Abu l-Ḥasan. *The Arithmetic of al-Uqlīdisī* (transl. and comm. A. S. Saidan). Dordrecht/Boston: Reidel, 1978.

Van der Waerden, B. L. and M. Folkerts. *Written Numerals.* Walton Hall, UK: The Open University Press, 1976.

Chapter 3

Geometrical Constructions in the Islamic World

§1. Euclidean Constructions

That geometrical constructions were of keen interest to the ancient Greek geometers is evident from the fact that Euclid devoted two of the thirteen books of his *Elements* to an account of some of the constructions that had been done up to his time. In Book IV, Euclid explains how to construct an equilateral triangle, a square and the regular pentagon, hexagon, octagon, decagon and 15-gon. In Book XIII he tells how to construct the regular polyhedra, namely the tetrahedron, cube, octahedron, dodecahedron and icosahedron—which have, respectively, 4, 6, 8, 12 and 20 faces.

Much of the Euclidean geometry studied in our schools deals with the geometry of triangles and circles, where the instruments used for construction are the straightedge and the compass. The former is used to join two points with a straight-line segment or to extend a straight-line segment in either direction, and the latter is used to draw a circle with an arbitrary point as center and passing through any given point. The ability to join points or to extend straight-line segments is assumed in Postulates 1 and 2 of *The Elements*, and the ability to draw arbitrary circles is postulated in Postulate 3. Perhaps for these reasons the straightedge and compass are called "the Euclidean tools".

The student should be careful to distinguish between a straightedge and a ruler, for, unlike the ruler, the straightedge is not assumed to have parallel edges or any marks along an edge. Similarly, the compass Euclid assumes for drawing circles is not the rigid compass we are used to, which stays set at whatever distance we open it to and may be used for transferring lengths. It is, rather, a compass that, once set on the paper, will draw around a given point as center the circle that passes through any other point, but it will not transfer lengths. For this reason it has been called a collapsible compass, since its legs fall back together when they are removed from the drawing plane.

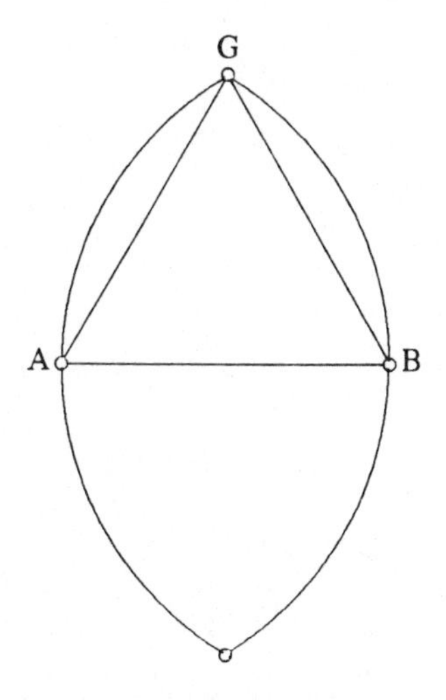

Fig. 3.1

The first thing Euclid does in *The Elements* is to show that his collapsible compasses are also able to transfer lengths, and to introduce this chapter on geometrical constructions we shall give the essentials of Euclid's proof. It is the first systematic discussion of a theory of constructions, in that it shows how constructions with one kind of tool can be done with a (seemingly) weaker tool.

In Proposition 1 of Book I, Euclid solved the following problem: On a given segment AB construct an equilateral triangle ABG (Fig. 3.1). By the properties of his compass Euclid is allowed to construct the circles centered at A and B having radii AB. If they intersect at G then $AG = AB = BG$, so that $\Delta(ABG)$ is equilateral.

In Prop. 2 Euclid shows how to place at a given point D a straight-line segment DW equal to a given segment AB (Fig. 3.2). He says to draw the straight line BD and, by Prop. 1, to construct the equilateral triangle BDG. Draw the circle with center B and radius BA, and let this circle cut GB, extended, at E. Finally, let the circle with center G and radius GE cut the side GD extended at W. Then $DW = GW - GD = GE - GB = BE = AB$, which was required.

Having proved these two propositions, Euclid finishes the demonstration with Prop. 3, where he shows how to cut off at the point Z on a given segment DE a segment equal to a given segment AB (Fig. 3.3). By Prop. 2 he can construct a segment ZF equal to AB, and the circle with center Z and radius ZF will intersect the line segment DE (extended if necessary) at a point G. The radius ZG is equal to AB. With this proposition Euclid is able to transfer a given length from any point to any other point, and therefore has shown that the collapsible compass is able to do the same operations as ours.

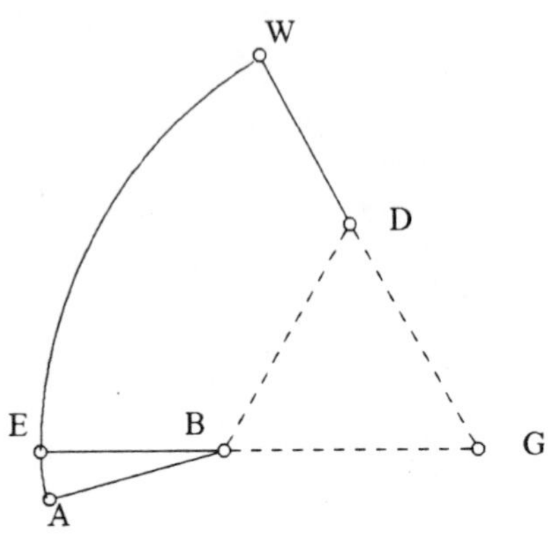

Fig. 3.2

The reader has doubtless observed that this demonstration serves no practical need. The Greeks knew as well as we how to lock two compass arms in a fixed position, so why did Euclid bother with some less powerful tool? The answer lies in the fact that one of Euclid's concerns was with achieving the greatest possible economy of basic assumptions. This concern is a matter of taste rather than of logical necessity, but it has been characteristic of mathematicians since the time of the Greek geometers. We shall see that it was a concern that the geometers of Islam shared.

Other basic constructions in *The Elements* are those of the perpendicular to a given line passing through a given point, the division of a line segment into an arbitrary number of equal segments, the bisector of an angle and the tangent to a given circle passing through a given point outside the circle, and these are probably familiar to the reader.

§2. Greek Sources for Islamic Geometry

Writers of the Islamic world were familiar with *The Elements* since the late eighth century through the translations that were done in Baghdad on commission for the caliphs Hārūn al-Rashīd and al-Ma'mūn. The numerous Arabic editions and commentaries which have survived since that time testify to the immense influence of Euclid's *Elements* on Islamic mathematics, since it was one of the basic texts that any student of mathematics and astronomy would have to read.

The name of Euclid, however, had to share honors in Islamic mathematics with that of another mathematician, namely Archimedes, whose treatise *On the Sphere and Cylinder* excited great admiration among the Muslim mathematicians and inspired some of their best works. In the preface to his book Archimedes mentions his discovery of the area of the segment of a parabola

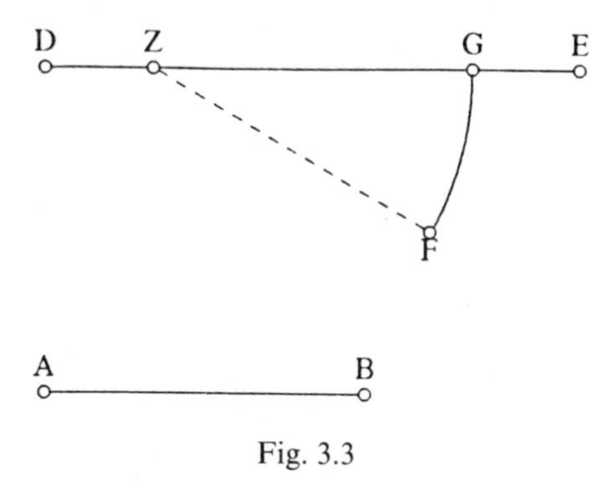

Fig. 3.3

and, since his treatise on the subject was not known in the Muslim world, this reference stimulated Thābit b. Qurra and his grandson Ibrāhīm b. Sinān to a successful search for proofs of Archimedes' result. In addition, the problem posed in the second part of Archimedes' work, that of dividing a sphere into two segments by a plane so that the two segments have to one another a given ratio, acted as a considerable stimulus to investigations in algebra and the conic sections. Another treatise, attributed to Archimedes by Arabic sources, but unknown in Greek, is titled *The Heptagon in the Circle*. It was put into Arabic by Thābit b. Qurra, who translated or revised translations of all Archimedean works extant in medieval Arabic. This work takes up the problem of constructing a regular polygon not discussed by Euclid, namely, the seven-sided polygon, the first unsolved case after Euclid's construction of polygons with 3, 4, 5 and 6 sides. The extensive literature on Archimedes' works is indicative of the fact that they formed the second pillar of Islamic geometry.

The third Grecian column supporting geometrical research in Islam was Apollonios of Perga's *The Conics*, a work in eight books (chapters) written around 200 B.C. Apollonios' work is difficult, and in the last few books he treats advanced topics such as the minimum distance from a point to a conic, so it is no surprise that only the first four of the original eight books survive in Greek. Evidently the remaining books were either too specialized or too difficult for the scholars of late antiquity. It is a testimony to the ability of the Muslim geometers that seven books survive in Arabic, and the bibliographer al-Nadīm tells us that in the tenth century parts of the eighth book were also extant. The work not only formed a base for much advanced research in geometry and optics—and even algebra, as we shall see in the work of ʿUmar Khayyām—but also it inspired one of the most able of Muslim scientists, Ibn al-Haytham, to attempt a restoration of the eighth book.

Although the student meets the basic Euclidean constructions during the high-school years, the elementary properties of the conic sections are often not encountered until a first university course in the calculus, and then by an approach that is quite different from that used by the Muslim authors. For these reasons, an account of some of the basic ideas in Apollonios' book will provide some necessary background for understanding the material in the following sections.

§3. Apollonios' Theory of the Conics

A surface of a double cone is formed by the straight lines that pass through the points on the circumference of a circle, called a *base*, and a fixed point not in the plane of the base (Fig. 3.4). Any one of the straight lines is called a *generator* of the surface, the fixed point is its *vertex* and the straight line through the vertex and the center of the base is called the *axis*. A *cone* is the solid figure bounded by the part of the surface of a double cone between the vertex and a base.

Euclid and Archimedes both wrote about conic sections before Apollonios, but in their treatments of conic sections the cone was the so-called right cone, in which the axis is perpendicular to the base circle. The right cone was then cut by a plane perpendicular to a generator, and in this way

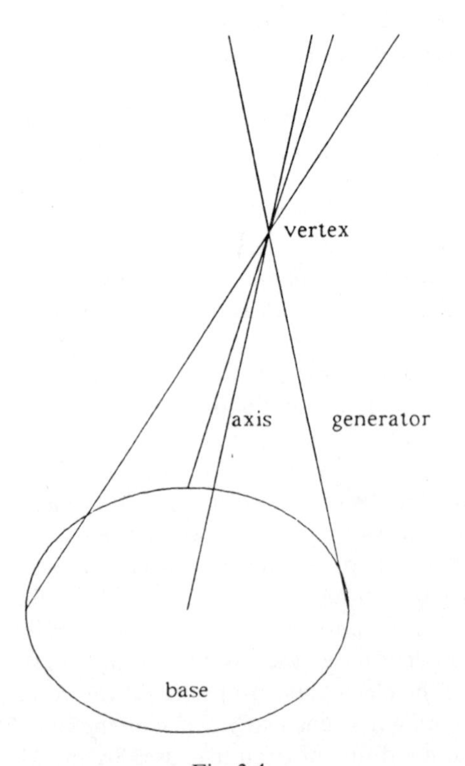

Fig. 3.4

one obtained a plane section, the kind of section depending on the angle at the vertex of the cone. Thus in the ancient world the conic sections were plane figures, whereas we look at the boundaries of these plane figures and think of the conic sections as curves.

Apollonios generalized this method of generating the conic sections by considering plane sections of an arbitrary double cone, whose axis may be skew to the base, and he showed that, apart from the circle, only the three known conic sections could arise.

At the beginning of his *Conics* Apollonios used the fact that these figures are sections of cones only to establish their basic properties, called "symptoms", and for the rest of the eight books he proved everything from these symptoms. Since, in the following, we shall be concerned only with parabola and hyperbola we shall limit our discussion of the symptoms to these two sections.

According to Apollonios a parabola is the common section of a cone and a plane when the plane is parallel to a generator of the cone and the hyperbola is either one of the two common sections formed when the plane meets both parts of the double cone. In either section any line joining two points on the boundary is called a chord. Apollonios showed that the midpoints of all chords parallel to a fixed chord lie on a straight line, and, if this straight line intersects the boundary in A, then the tangent at A is parallel to all the chords. The straight line is called a *diameter* of the section and an intersection of a diameter with the boundary is called a *vertex* of the section. The half-chords lying on one side of the diameter are called the *ordinates* to the diameter. When the ordinates are perpendicular to the diameter, such a diameter is unique and is called the *axis*. We have illustrated these ideas for a parabola in Fig. 3.5, where FE is a chord, YZ and UV are parallel to FE, and AB is the diameter passing through the midpoints of these chords. XY is a typical ordinate to the diameter AB, and the line CD is the axis.

Symptom of the Parabola

In the case of the parabola, the diameters are all parallel to the axis CD. Let AB be a given diameter, X be any point on AB, and XY the ordinate at X. Apollonios showed (Fig. 3.5) that corresponding to the diameter AB there is a fixed line segment p so that the rectangle that is equal to the square on XY and has one of its sides equal to AX will have its other side exactly fit the segment p. The segment p is called the parameter (or *latus rectum*) belonging to diameter AB. If we set $AX = x$ and $XY = y$, then the Apollonian symptom becomes the modern equation $p \cdot x = y^2$. The Greek term Apollonios used to describe this is *paraballetai* and so he called this section *parabolē*, from which comes our word "parabola". (The word *paraballetai* literally means "it is put alongside", referring to the rectangle exactly fitting along the parameter.)

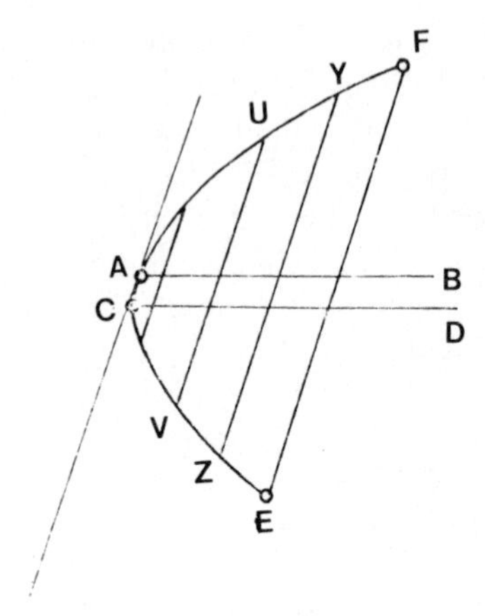

Fig. 3.5

Symptom of the Hyperbola

Here the curve has a center, the point on the axis midway between the vertices of the two sections (Fig. 3.6). Any line through this center is a diameter and the center bisects the part of a diameter between the two branches. Let C, C′ be the endpoints of the part of some diameter between the two branches of the curve and let $a = CC'$, called the transverse side (or *latus transversum*). Apollonios proved that, corresponding to a there is a segment p with the following property: A rectangle that is equal to the square on an ordinate XY and has one side equal to CX will have its other side exceed p. Moreover, the rectangle (shaded in Fig. 3.6) contained by the excess of this side over p and by CX is similar to the rectangle whose sides are a and p. Thus the other side s satisfies the proportion $s : CX = p : a$, i.e. $s = (p/a) \cdot CX$.

To see what the symptom for the hyperbola says geometrically let C′ be the other end of the diameter and let CP be the parameter. Also let the perpendicular to CX at X meet the line C′P at E. Then Apollonios' symptom implies that the rectangle whose sides are CX and XE is equal to $(XY)^2$. (The side p is called the parameter.) Since the Greek word for "it exceeds" is "hyperballetai", Appolonios' name for this case is *hyperbolē*, whence comes our word "hyperbola". Again, if we let $CX = x$ and $XY = y$ the symptom becomes

$$y^2 = (p + s)x = px + (p/a)x^2,$$

which is a modern equation for the hyperbola.

The above properties of the sections were hardly Apollonios' discoveries, since they were known to Archimedes, but one of Apollonios' contributions was, having shown that the symptoms characterize the conic sections, to use the properties stated in the symptoms to name the sections parabola, hyperbola and ellipse ("ellipse" from *elleipsis*, meaning "falling short").

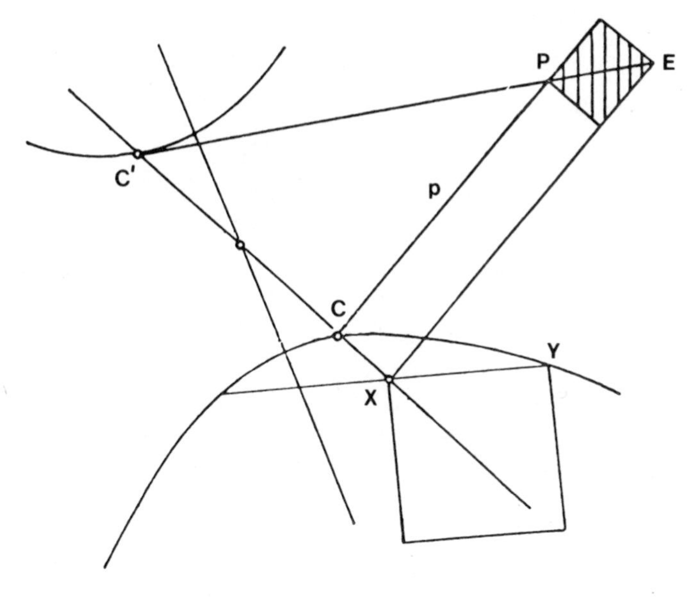

Fig. 3.6

§4. Abū Sahl on the Regular Heptagon

Archimedes' Construction of the Regular Heptagon

In both the Greek and the Islamic worlds the principal uses of the conic sections (other than the circle) were in geometrical constructions, the design of mirrors that focus light to cause burning, and the theory of sundials. The use of ellipses in astronomy to model the planetary orbits was introduced by J. Kepler early in the seventeenth century. As for conics' use in geometrical constructions, it was Menaechmos who, in the first half of the fourth century B.C., invented the conic sections and used them to solve the problem of constructing a cube whose volume is twice that of a given cube. If the side of the given cube is a and the side of the required cube is b then $b^3 = 2a^3$, so $b = \sqrt[3]{2}\, a$, and the problem is one of scaling up a line segment of length a to one of length $\sqrt[3]{2}\, a$. (Because of work on the theory of equations by the nineteenth century French mathematician E. Galois we know that this cannot be done with straightedge and compass. However, both the ancient Greek and the Muslim geometers realized that this construction and many others are possible with conic sections.)

Despite this, one construction remained anomalous and unexplained. This is a construction of the regular heptagon which is attributed to Archimedes. So puzzling was it that the tenth-century Muslim mathematician Abu l-Jūd remarked with some justice that "perhaps its execution is more difficult and its proof more remote than that for which it serves as a premiss".

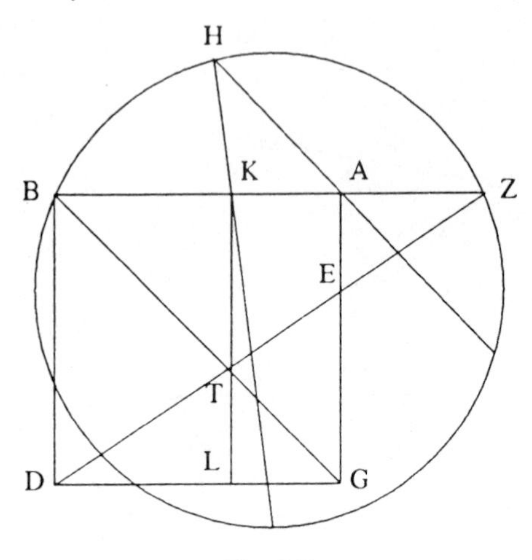

Fig. 3.7

Archimedes begins with the square ABDG and its diagonal BG (Fig. 3.7). He now turns a straightedge around D so that it crosses the diagonal BG, the side AG and the side BA extended at points T, E and Z respectively, and so that $\triangle(AEZ)$ has the same area as $\triangle(DTG)$. Lastly he draws KTL parallel to AG. He then proves that K and A divide the segment BZ so that the three segments BK, KA and AZ can form a triangle and so that $BA \cdot BK = ZA^2$ and $KZ \cdot KA = KB^2$. Thus, form $\triangle(KHA)$ so that $KH = KB$ and $AH = AZ$, and through B, H and Z draw the circle BHZ. Archimedes proves that $\overset{\frown}{BH}$ is one-seventh of the circumference of the circle.

No other construction like this one is known in Greek or Islamic mathematics, and this uniqueness is a hallmark of Archimedes' work. However, for all its elegance, it must be admitted that, as Abu l-Jūd suggests, it raises as many problems as it solves. Of course, if one thinks of a straightedge rotating around D so its passes between A and G then, as it moves toward A, $\triangle(AEZ)$ can become arbitrarily small while $\triangle(DTG)$ approaches one-quarter of the square. On the other hand, as the straightedge approaches G, $\triangle(AEZ)$ becomes arbitrarily large and $\triangle(DTG)$ becomes arbitrarily small. Thus, for some intermediate position, the two triangles will be equal, and so Archimedes' procedure is, if we like, an existence proof, but it is hardly a construction. Thus the problem remained one not constructively solved for almost 1200 years.

Abū Sahl's Analysis

In the latter half of the tenth century, however, in Baghdad and the surrounding area, there gathered a group of remarkable scientists from all over the eastern part of the Islamic world under the patronage of a series of kings belonging the the Būyid family. Foremost of these kings was ʿAḍud al-Daula ("Arm of the State"), and one of the chief scientists at his court was Abū Sahl al-Kūhī, who came from the mountainous region ("kūh" is the Persian word for "mountain") south of the Caspian Sea. According to the biographer al-Bayhaqī, who lived about a century later, Abū Sahl was originally a juggler of glass bottles in the market place of Baghdad, but then he gave up juggling for study and research in the sciences. Perhaps it was his experience as a juggler that aroused his interest in centers of gravity, for correspondence of his contains some of the deepest theorems on centers of gravity since the time of Archimedes. In fact, Abū Sahl knew Archimedes' work well, and he wrote a commentary on Book II of *On the Sphere and Cylinder*, in which he explained how to solve by conic sections the problem of constructing a sphere with a segment similar to a segment of one sphere and having surface area equal to a segment of a second sphere. In addition, he wrote a work on "the complete compass", an instrument for drawing conic sections. Consistently with this interest and experience in conic sections, Abū Sahl looked at the problem of constructing a regular heptagon, and he saw that a solution lay in conic sections. His method of attack was inspired by Archimedes' proof, and when he refers to the construction of the heptagon as a problem that no geometer before him, "not even Archimedes", had been able to solve, he was no doubt referring to the problem of actually doing the construction that Archimedes' method calls for.

Abū Sahl's method is to *analyze* the problem first, that is to suppose that the heptagon has been constructed and to reason backward, by a series of inferences that can be validly reversed. Analysis is an ancient method, which Proklos, a biased source, attributes to Plato. According to this method the mathematician assumes what is to be proved and then reasons from this until he reaches the given. If the chain of reasoning can be reversed then he has found the *synthesis*, or proof, of what is required starting from the given. Abū Sahl used analysis to find a series of constructions equivalent to that of the regular heptagon, until he arrived at given. Many geometers of the late tenth century felt that a complete solution of a problem required both the analysis and the synthesis, and Ibrāhīm b. Sinān, whom we shall speak of later in this chapter, wrote a treatise on these two methods.

We shall, however, present only the analysis as it is found in a treatise written by Abū Sahl and dedicated to King ʿAḍud al-Daula, that is we shall trace the series of constructions by which Abū Sahl reduced the problem of constructing the regular heptagon to one of constructing two conic sections. When he has done this he has shown how a peculiar construction, fitting into no theory, could be fitted into the theory of conic sections. Such a unification of disparate mathematical methods is the very stuff of which mathematical progress is made.

First Reduction: From Heptagon to Triangle

Suppose that in the circle ABG we have succeeded in constructing the side $\overset{\frown}{BG}$ of a regular heptagon (Fig. 3.8) and that $\overset{\frown}{AB} = 2\overset{\frown}{BG}$. Then arc

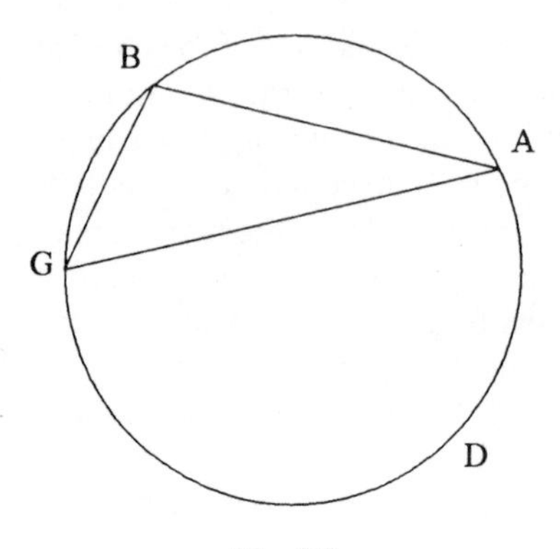

Fig. 3.8

$\widehat{ABG} = 3\widehat{BG}$ and, since $\widehat{BG}$ is one-seventh of the whole circumference, $\widehat{ADG} = 4BG$. According to VI,33 of Euclid's *Elements* angles of $\triangle(ABG)$ on the circumference are in the same proportion as the arcs they subtend, and therefore $\sphericalangle B = 4\sphericalangle A$ while $\sphericalangle G = 2\sphericalangle A$. Thus, the construction is reduced to the problem of constructing a triangle whose angles are in the ratio 4:2:1.

Second Reduction: From Triangle to Division of Line Segment

Let ABG be a triangle so that $\sphericalangle B = 2\sphericalangle G = 4\sphericalangle A$ (Fig. 3.9) and prolong BG in both directions to D and E so that DG = GA and EB = BA. Complete $\triangle(AED)$. (In the following $\sphericalangle A$, $\sphericalangle B$, $\sphericalangle G$ will denote the angles of $\triangle(ABG)$ at the corresponding vertices, and other angles at these vertices will be referred to unambiguously.) The idea of the proof is to show that $\sphericalangle A = \sphericalangle D$ so that the two triangles ABG and DBA are similar, then to show that $\sphericalangle BAE = \sphericalangle G$ so that the two triangles AEB and GEA are similar. When this is done then, by the first similarity, DB/BA = AB/BG and, by the second, GE/AE = AE/BE. Thus it follows that

$$BA^2 = DB \cdot BG \quad \text{and} \quad EA^2 = GE \cdot EB.$$

However, since AB = BE, $\sphericalangle E = \sphericalangle BAE = \sphericalangle G$ so that EA = AG = GD. Thus, the second of the preceding equalities becomes $GD^2 = GE \cdot EB$ and

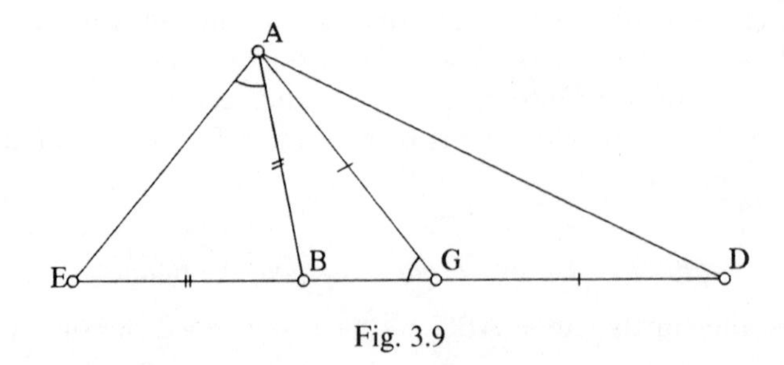

Fig. 3.9

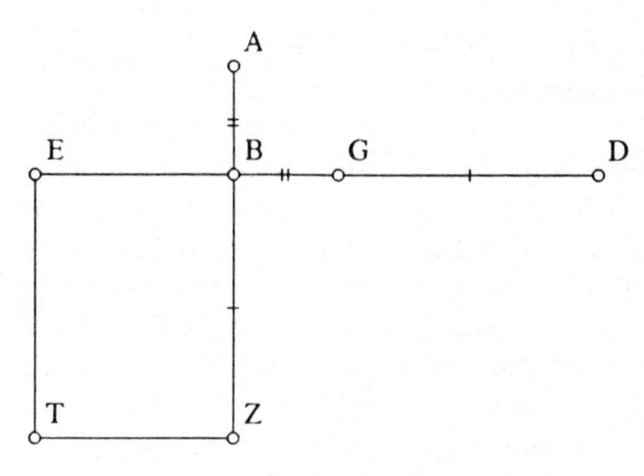

Fig. 3.10

the first becomes $BE^2 = DB \cdot BG$ since $BA = BE$. Thus, once we have shown that $\sphericalangle A = \sphericalangle D$ and $\sphericalangle BAE = \sphericalangle G$ we will have shown that the construction of a regular heptagon implies the division of a straight line ED at two points B, G so that

$$(1)\ GE \cdot EB = GD^2 \quad \text{and} \quad (2)\ DB \cdot BG = BE^2.$$

As for the angles, then, notice that $\sphericalangle BGA$ is an exterior angle of the isosceles triangle AGB, where $AG = GD$, so that $\sphericalangle BGA = \sphericalangle DAG + \sphericalangle D = 2\sphericalangle D$. But we are given that $\sphericalangle BGA = 2\sphericalangle A$, so that $\sphericalangle A = \sphericalangle D$. In the case of the other angle, observe that $\sphericalangle B$ is an exterior angle of the isosceles triangle ABE so that $\sphericalangle B = 2\sphericalangle BAE$ while, at the same time, $\sphericalangle B = 2\sphericalangle G$, so that $\sphericalangle BAE = \sphericalangle G$.

Third Reduction: *From the Divided Line Segment to Conic Sections*

Let ED be a line segment divided at B, G so that (1) and (2) above are satisfied (Fig. 3.10). Draw ABZ perpendicular to ED with $AB = BG$ and $BZ = GD$, and then complete the rectangle BZTE. Then $ZA \cdot AB = DB \cdot BG = BE^2$, and since $AB = BG$ and $BE = TZ$ we may write $ZA \cdot BG = TZ^2$. which says that the point T lies on a parabola whose vertex is A and whose parameter is BG.

On the other hand, by (1) $GE \cdot EB = GD^2$; but, $GD = BZ = ET$, so $GE \cdot EB = ET^2$ so that T lies on a hyperbola with vertex B whose transverse side and parameter are both equal to the segment BG.

Our analysis has now led us to two conics—a parabola and a hyperbola—both determined by the division of ED at B, G. The intersection point, T, of these two conics determines the lengths ET and TZ, and these produce the remaining two segments $GD = ET$ and $EB = TZ$ with the property that the line EBGD is divided at B and G so that (1) and (2) are satisfied. Thus given BG, the side of the heptagon we wish to construct, we may construct the segment EBGD, then the $\triangle(ABG)$, and finally the heptagon. Of course, once a heptagon has been constructed in some circle it may be constructed in any other circle by similarity.

Abū Sahl was under no illusions that the conic sections could be constructed by straightedge and compass. As we have mentioned, he wrote a special treatise to describe an instrument, the complete compass, that could draw conic sections. Rather, the point of this treatise is that if one is given the next class of curves beyond a straight line and circle, namely the conic sections, then one may construct in any circle the side of a regular heptagon. Many centuries after Abū Sahl's time mathematicians, following Descartes' discovery of analytic geometry, would begin to classify curves according to the degree of the algebraic expression for the curves—thus "quadratic", "cubic", etc. In the ancient world, however, problems were described as *plane*, *solid* or *curvilinear* according as it was possible to solve them by straight lines or circles, sections of cones, or in more complicated ways. In this context Abū Sahl's proof may be seen as showing that the construction of a regular heptagon belongs to an intermediate class of problems whose solutions demand at worst *solid* curves. Thus, he limits both the level of difficulty of the problem and the means necessary to solve it, and he places the problem within the context of the known mathematical theory of conic sections.

§5. The Construction of the Regular Nonagon

Verging Constructions

The construction of the regular nonagon, i.e. the regular polygon with nine equal sides, is a special case of the trisection of the angle, since the central angle of a nonagon is $360°/9 = 120°/3$ (Fig. 3.11). But $120°$ is the central angle subtending the side of an equilateral triangle inscribed in a circle, so the regular nonagon in a circle can be constructed by trisecting this

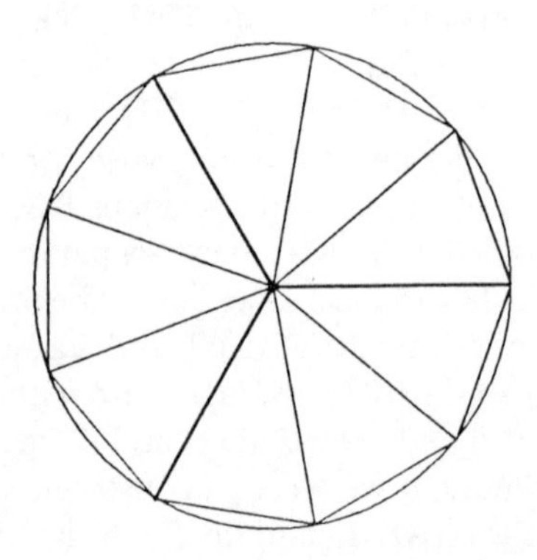

Fig. 3.11

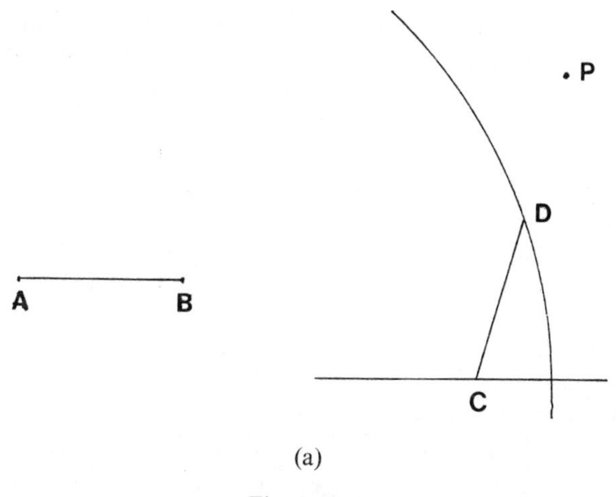

(a)

Fig. 3.12(a)

angle. This was well known to the ancient Greeks, and Pappos of Alexandria gives three methods of trisecting an angle, all of which use conic sections. The only ancient method that seems to have been transmitted to the Muslim scholars may be found in the works of Thābit b. Qurra and his patron and colleague Aḥmad b. Shākir. However, the Greek treatise from which these two worked seems to have been lost, for near the end of the tenth century the geometer ʿAbd al-Jalīl al-Sijzī, who, as a younger contemporary of Abū Sahl al-Kūhī, participated in solar observations with him at Shirāz in 969–970, wrote, "It was not possible for any of the ancients to solve this problem (trisecting the angle) despite their strong desire . . . ." What al-Sijzī meant by this becomes clear later on when he refers to "Another lemma of one of the ancients (for trisecting an angle) that uses a ruler and moving geometry (i.e. verging constructions), but which we have to solve by fixed geometry". Thus al-Sijzī knew of ancient procedures for trisecting the angle, but they were of a kind he called "moving geometry". What al-Sijzī was referring to by the phrase "moving geometry" is a kind of construction that Apollonios and other Greek writers called *vergings*. In a verging construction, one is given two curves, usually straight lines or circular arcs, point P not on the curves and also a straight line *segment* AB. The problem is to construct a straight-line segment CD = AB so that one endpoint lies on one of the given curves, the other endpoint on the other, and so that CD verges toward P, that is, when extended it passes through P (Fig. 3.12(a)).

A verging construction was used by Hippocrates of Chios in the early fourth century B.C. to find the area of lunules, and later Archimedes used vergings to prove theorems on the spiral. There are as well, in Arabic manuscripts ascribed to Archimedes, uses of vergings to trisect an angle and to construct the regular heptagon—and all of these are done without further comment. This may be because he felt that such constructions were as legitimate as any others. In any case, geometers well before his time knew how to perform such constructions by means of the conic sections, and Apollonios, a younger contemporary of Archimedes, wrote a two-part book on vergings that could be done by a straightedge and compass alone.

Fixed Versus Moving Geometry

By the tenth century, however, some geometers felt that verging constructions were not acceptable as independent operations and they tried to find other solutions to problems their predecessors had solved by vergings. The solution al-Sijzī offers to the problem of trisecting the angle is one that seems to have originated with the Muslim geometers. Al-Sijzī refers to its main lemma as "the lemma of Abū Sahl al-Kūhī"; furthermore, the whole trisection appears in Abū Sahl's work, so it seems fairly certain that this trisection is one more discovery of the gifted geometer from the region south of the Caspian Sea. Since Abū Sahl was at Shirāz with al-Sijzī 969–970 it is possible that al-Sijzī learned of the trisection then.

Abū Sahl's Trisection of the Angle

Trisecting the angle according to Abū Sahl's method depends on solving the following problem: Let the semicircle AZD be given and let AD be its diameter, with H its center (Fig. 3.12(b)). Let the angle ABG also be given. It is required to find on the diameter a point E so that if EZ is parallel to BG then $EZ^2 = EH \cdot ED$.

Construction. To do this we construct on AH as diameter the hyperbola HZL, whose parameter and transverse side are both equal to AH and so that the angle the ordinate makes with the diameter is equal to $\measuredangle ABG$. (This construction was explained by Apollonios in *The Conics*, Book I, Props. 54 and 55, and we shall later on see a practical method due to Ibrāhīm ibn Sinān for constructing such a hyperbola.) Let this hyperbola cut the semicircle at Z,

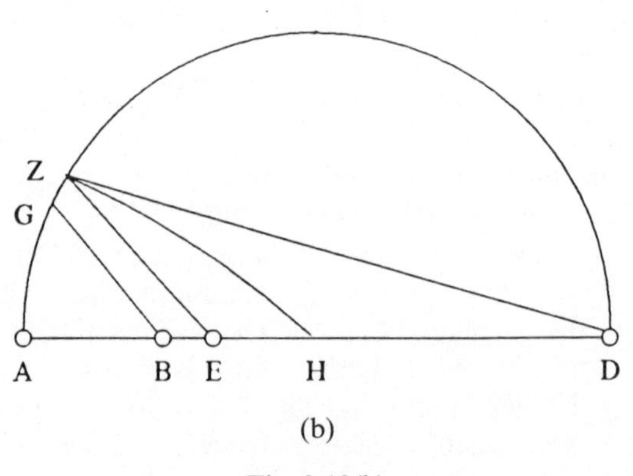

(b)

Fig. 3.12(b)

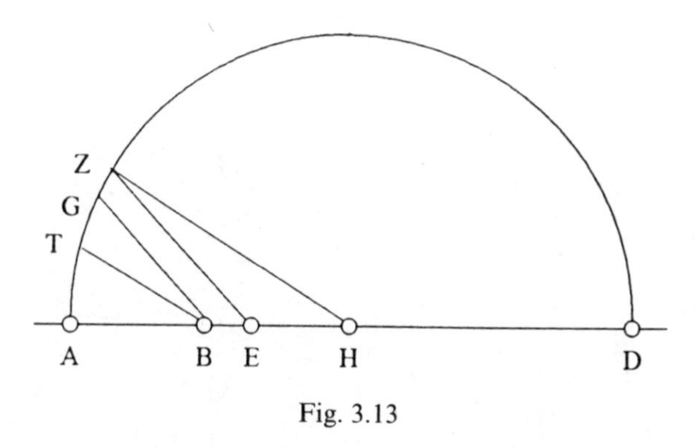

Fig. 3.13

and draw EZ and BG. Then, by the properties of the hyperbola,

$$EH \cdot ED/EZ^2 = \text{Transverse side}(=AH)/\text{Parameter}(=AH),$$

$$\text{so that} \quad EZ^2 = EH \cdot ED,$$

and the problem is solved.

Now to trisect an arbitrary acute angle ABG according to al-Kūhī's method extend the side AB in a straight line to D, where the length BD may be chosen arbitrarily (Fig. 3.13). On AD as diameter draw the semicircle AGZD and let H be the midpoint of AD. Draw EZ parallel to BG, where E is chosen, as above, so that $EH \cdot ED = EZ^2$. Then draw ZH and ZD, as well as BT parallel to ZH. Then $\measuredangle ABT = 2\measuredangle TBG$. Then $\measuredangle ABG = \measuredangle ABT + \measuredangle TBG = 2\measuredangle TBG + \measuredangle TBG = 3\measuredangle TBG$, so we have trisected the given angle ABG.)

To prove the foregoing, notice that the condition $EH \cdot ED = EZ^2$, which the point E satisfies, implies that $ED/EZ = EZ/EH$, and since the two triangles HEZ, ZED have the angle E in common and the pair of sides containing this angle proportional it follows that they are similar, so $\measuredangle EZH = \measuredangle D$. However, the two sides ZH, HD of $\Delta(ZHD)$ are equal, so that $\measuredangle HZD = \measuredangle D$, and thus $\measuredangle EZH = \measuredangle HZD$. Now $\measuredangle EHZ$ is an exterior angle of the $\Delta(ZHD)$, so that $\measuredangle EHZ = \measuredangle HZD + \measuredangle D = 2\measuredangle HZD$. However, since $\measuredangle BEZ$ is exterior to the $\Delta(ZEH)$, it follows that $\measuredangle BEZ = \measuredangle EZH + \measuredangle EHZ = \measuredangle HZD + 2\measuredangle HZD = 3\measuredangle HZD$. Thus, $\measuredangle EHZ = (2/3)\ \measuredangle ABG$, and since BT is parallel to ZH, $\measuredangle ABT = (2/3)\ \measuredangle ABG$, i,e. $\measuredangle TBG = (1/3)\ \measuredangle ABG$.

§6. Construction of the Conic Sections

Life of Ibrāhīm b. Sinān

The conic sections were not used for theoretical purposes only. Indeed, the Greeks had realized that as the sun traces its circular path across the sky during the day, the rays that pass over the tip of a vertical rod set in the earth

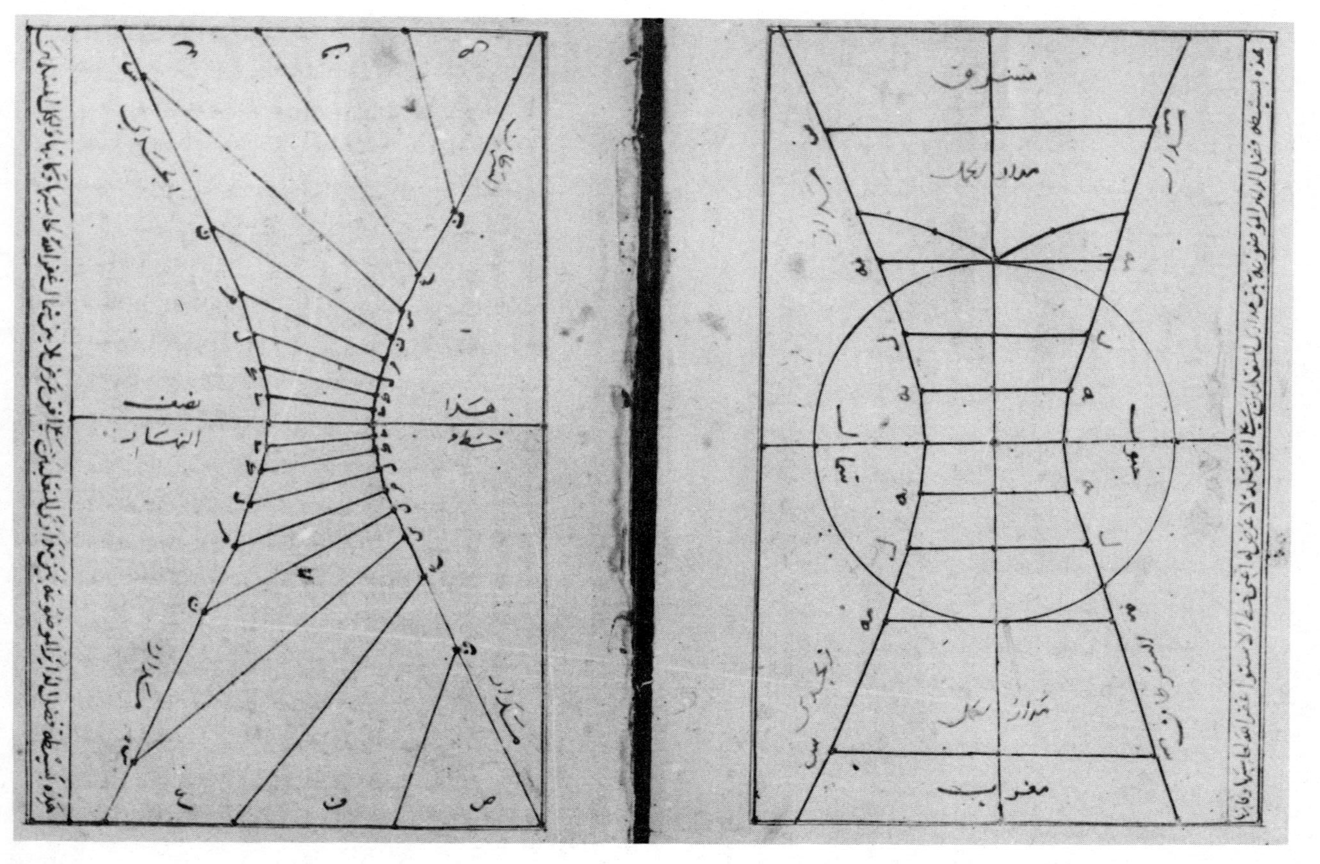

Plate 3.1. Diagrams of a sundial for the latitude of Cairo and a universal sundial in a fifteenth century Egyptian treatise on sundial theory by the *muwaqqit* al-Karādīsī. (A *muwaqqit* is a person who determines the times of prayer in Islam.) The two hyperbolas represent the paths of the shadow of the pointer (*miqyās*) at the two solstices. (Taken from MS Cairo Dār al-kutub riyāḍa 892. Courtesy of the Egyptian National Library.)

form a double cone, and, because the plane of the horizon cuts both parts of this cone, the section of the cone by the horizon plane must be a hyperbola on a horizontal surface (see Plate 3.1). It was therefore of use to instrument-makers to know how to construct hyperbolas, since it would be necessary to engrave or cut them on the surfaces of sundials. No doubt there were for this purpose "tricks of the trade" that craftsmen used, and, perhaps, few artisans ever looked into a book explaining how to draw a hyperbola. However, whatever may have been the relationship of theory to practice, such treatises were written, one of them by the grandson of Thābit b. Qurra, named Ibrāhīm b. Sinān. Although his life ended by a liver tumor in A.D. 946 at the age of 37 his surviving works ensure his reputation as an important figure in the history of mathematics. His treatment of the area of a segment of a parabola is the simplest that has come down to us from the period prior to the Renaissance. (He tells us he invented the proof in order to rescue the family's scientific reputation when he heard accusations that his grandfather's method was too long-winded.) One of his unstudied works we have already referred to, namely the one titled *On the Method of Analysis and Synthesis in Geometrical Problems*, so he was yet another of the tenth-century scientists who were concerned not only with particular problems but with whole methods or theories. In addition, in his work on sundials he treats the design of all possible kinds of dials according to a single, unified procedure, and it represents a fresh, successful attack on problems that had often defeated his predecessors.

In this section we shall focus on another work of his, *On Drawing the Three Conic Sections*. This work contains a careful discussion, with proofs, of how to draw the parabola and ellipse, as well as three methods for drawing the hyperbola. Perhaps so many methods were given for the hyperbola because it was the one of most interest to the instrument-makers. From this work we shall present two selections, the one dealing with the construction of the parabola, which is necessary for the construction of burning mirrors, and the other giving one of three methods for drawing the hyperbola.

Ibrāhīm b. Sinān on the Parabola

Ibrāhīm's method is the following. On the line AG (Fig. 3.14) mark off a fixed segment AB and construct BE perpendicular to AB. Now on BG pick as many points H, D, Z, ... as you wish. Starting with the point H, draw the semicircle whose diameter is AH, and let the perpendicular BE intersect it at T. Through T draw a line parallel to AB and through H draw a line parallel to BE. Let these lines intersect at K.

Next, draw a semicircle on AD as diameter and let it intersect BE at I. As before, draw lines through I and D parallel to AG and BE respectively, and let them meet at L. Do the same construction for the remaining points Z, ... to obtain corresponding points. Then the points B, K, L, M, ... lie on the

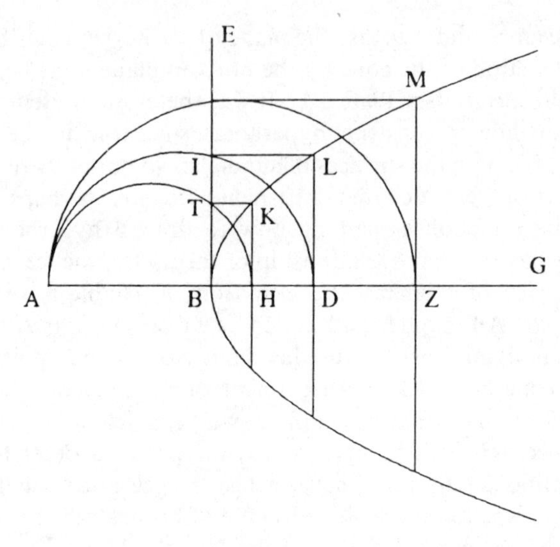

Fig. 3.14

parabola with vertex B, axis BG and parameter AB. If K′, L′, M′, . . . are chosen on the prolongations of KH, LD, MZ, . . . so that KH = HK′, LD = ĐŁ′, MZ = ZM′, . . . then they too lie on the parabola.

Ibrāhīm proves that K is on the parabola he has described as follows: If the parabola does not pass through K let it pass through another point on KH, say N. Then $NH^2 = AB \cdot BH$, by the property of the parabola. On the other hand, since TB is perpendicular to the diameter of the semicircle ATH, it follows from Euclid II,14 that $TB^2 = AB \cdot BH$. Further, by construction, TBHK is a parallelogram so TB = KH. Thus $KH^2 = TB^2 = AB \cdot BH = NH^2$, and so KH = NH, which is a contradiction. Thus K lies on the parabola we described, and the same proof, with the names of points changed, applies equally well to L, M, . . . and so the validity of the construction has been shown.

The reader who would like to try Ibrāhīm's method can make it easier by drawing the successive semicircles through A arbitrarily rather than taking their diameters as given in advance. This saves bisecting the lines AH, etc.

Ibrāhīm b. Sinān on the Hyperbola

As we have said, this is only one of the three methods he gives, but it is certainly the one most easily done. On a fixed segment AB (Fig. 3.15) draw a semicircle and prolong its diameter AB in the direction of B. On the half of the semicircle near B pick points as G, D, H, . . . and at each of these points construct the tangents to the semicircle GZ, DT, HI, ... Let these tangents meet the diameter extended at Z, T, I, . . . respectively and through these

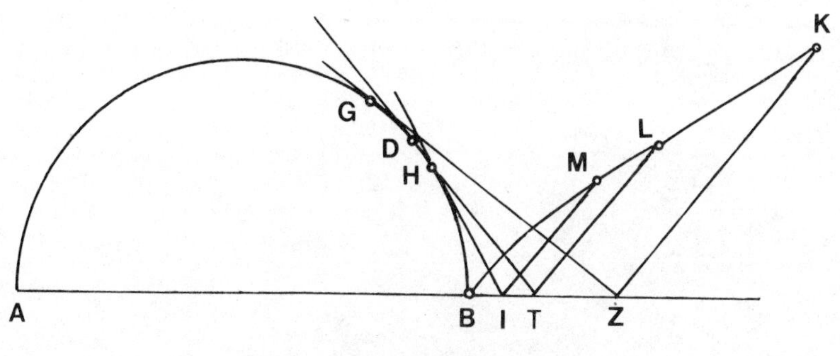

Fig. 3.15

points draw the parallel straight lines ZK, TL, IM, ... making an arbitrary angle with the line AB. On these lines, on the same side of AB, cut off ZK = GZ, TL = DT, IM = HI, ... Then the points K, L, M, ... lie on a hyperbola.

Indeed, since the lines GZ, DT, HI, ... are tangents to a circle it follows from Euclid III,36 that $GZ^2 = ZB \cdot ZA$, $DT^2 = TB \cdot TA$, $HI^2 = IB \cdot IA$, ... and since KZ = ZG, etc., it follows that $KZ^2 = ZB \cdot ZA$, $LT^2 = TB \cdot TA$ and $MI^2 = IB \cdot IA$. According to the symptom of the hyperbola, given earlier, these relations say that the points B, K, L, M, ... lie on a hyperbola with the line AB as a diameter, whose ordinates all make with the diameter angles equal to $\sphericalangle$KZG and whose parameter and transverse sides are both equal to AB. Here again, the rest of the one branch of the hyperbola can be constructed simply by prolonging DZ, LT, MI, ... an equal length beyond ABG to K′, L′, M′, ...

§7. The Islamic Dimension: Geometry with a Rusty Compass

An aspect of Islamic civilization that has always impressed outsiders has been the elaborate geometrical designs executed in wood, tile or mosaic and found in abundance throughout the Islamic world. For example, the especially fine regular tilings of the plane found in the Alhambra at Granada in Spain are admired the world over. A crafts tradition of such sophistication involves a considerable amount of geometrical knowledge, even if that knowledge was simply passed on from master to apprentice rather than being written down (see Plates 3.2 to 3.4).

Of course, there had been a strong tradition of geometrical design in the Middle East since the time of ancient Egypt, and this continued both in ancient Greece and elsewhere. At some time geometers became aware of this tradition and the problems the artisans solved, and they began to try to justify the procedures and to see how far various methods could be pushed.

Plate 3.2. This part of the facade of the Shir Dor madrasah in Samarqand illustrates the variety of elements to be found in Islamic art. It combines calligraphic elements (on the border and inside arch) with arabesques, geometric designs and both anthropomorphic and zoomorphic elements.

For example, in the Arabic version of the eighth book of Pappos of Alexandria's *Mathematical Collection* is a very interesting section on geometrical constructions that are possible using only a straightedge and a compass with one fixed opening, sometimes called a "rusty compass". Since the rest of the eighth book is devoted to instruments and machines of interest to people various crafts it seems likely that this section too was aimed at addressing problems encountered by craftsmen.

The continuing Islamic interest in these problems is witnessed by the fact that the Arabic text of Pappos' Book VIII was copied late in the tenth century by al-Sijzī, whom we have mentioned earlier in connection with the regular heptagon, from an earlier copy that belonged to the Bānū Mūsā, the patrons and friends of the ninth-century mathematician Thābit b. Qurra.

Another treatise on geometrical constructions with various restrictions on the tools is attributed to Abū Naṣr al-Farabī, who is best known today for his important commentaries on Aristotle and his great work on music. He was born in 870, when the Bānū Mūsā were old, and he taught philosophy in both Baghdad and Aleppo, an important trading city in northern Syria. He lived a long life of active scholarship and was killed by highway robbers

Plate 3.3. This detail from the facade of the Friday mosque in Isfahan combines pentagons, octagons and stellated decagons with nonconvex pieces to tile the plane. The pattern could obviously the extended infinitely in all directions.

Plate 3.4. An example from Isfahan of the kind of ceiling known as *al-muqarnar*. They were sufficiently common in Islamic architecture for al-Kāshī to devote a section of his *Calculators' Key* to their theory.

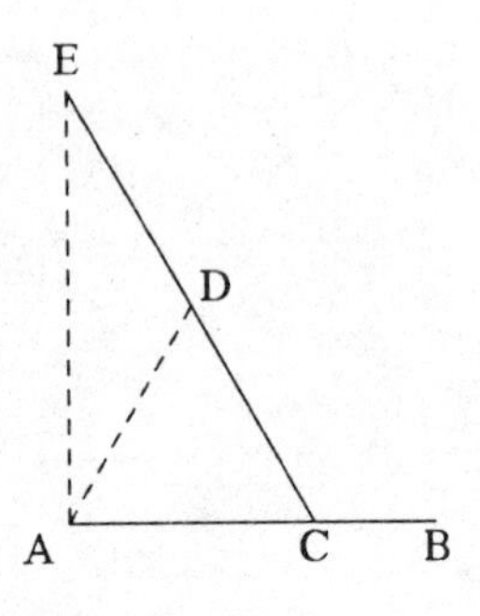

Fig. 3.16

outside Damascus in the year 950, not too long after Ibrāhīm b. Sinān died. In addition to his other work, he wrote a treatise with the title *A Book of Spiritual Crafts and Natural Secrets in the Details of Geometrical Figures.* Later, Abu l-Wafā', whom we shall speak of in the chapter on trigonometry, and who was young when al-Farabī died in 950, incorporated all of al-Farabī's work in his own, more prosaically-titled treatise, *On Those Parts of Geometry Needed by Craftsmen.* It is from this treatise that we have chosen the following extracts, where the numbering of the problems is our own.

Problem 1

To construct at the endpoint A of a segment AB a perpendicular to that segment, without prolonging the segment beyond A.

Procedure. *On* AB *mark off with the compass segment* AC (Fig. 3.16), *and, with the same opening, draw circles centered at* A *and* C, *which meet at* D. *Extend* CD *beyond* D *to* E *so that* ED = DC. *Then* ∡CAE *is a right angle.*

Proof. The circle that passes through E, A, C has D as a center since DC = DA = DE. Thus EC is a diameter of that circle and therefore ∡EAC is an angle in a semicircle and hence is a right angle.

Problem 2

To divide a line segment into any number of equal parts.

Procedure. *Let it be required to divide the line segment* AB (Fig. 3.17) *into the equal parts* AG = GD = DB. *At both endpoints erect perpendiculars* AE, BZ *in different directions and on them measure off equal segments* AH = HE = BT = TZ. *Join* H *to* Z *and* E *to* T *by straight lines which cut* AB *at* G, D, *respectively. Then* AG = GD = DB.

Proof. Indeed, AHG and BTD are two right triangles with equal angles at G and D (and therefore at H and T). In addition HA = BT. Thus the triangles are congruent and so AG = BD. Also the parallelism of HG and ED implies that the two triangles AHG and AED are similar, and thus DG/GA = EH/HA. But, EH = HA and so DG = GA.

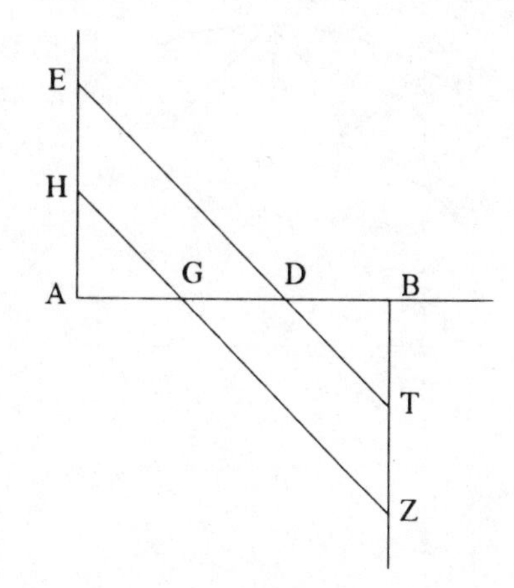

Fig. 3.17

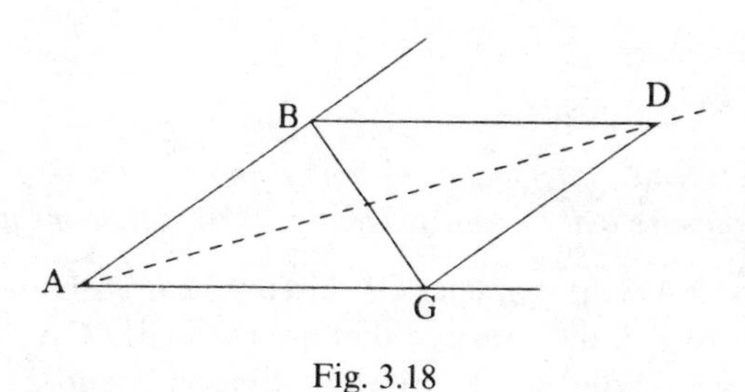

Fig. 3.18

Problem 3

To bisect a given angle, BAG.

Procedure. *The Euclidean method* (*Book* I,9) *involves cutting off equal segments* AB, AG *on the two sides of the angle, constructing the equilateral triangle on* BG, *and then joining* A, D *to bisect the angle. According to Abu l-Wafā'''s variation of this, the triangle* BGD *is isosceles with* BD = DG = AB, *the fixed compass opening* (Fig. 3.18).

Next Abu l-Wafā' finds the center of a given circle. We shall use this in explaining the next rusty compass construction, but we leave it for the reader to find the construction.

Problem 4

To construct a square in a given circle.

Procedure. *Locate the center S and draw a diameter* ASG (Fig. 3.19). *With compass opening equal to the radius* (*Exercise* 9) *mark off arcs* $\widehat{AZ}$, $\widehat{AE}$, $\widehat{GT}$ *and* $\widehat{GH}$ *and draw the lines* ZE *and* TH, *which cut the diameter at* I *and* K.

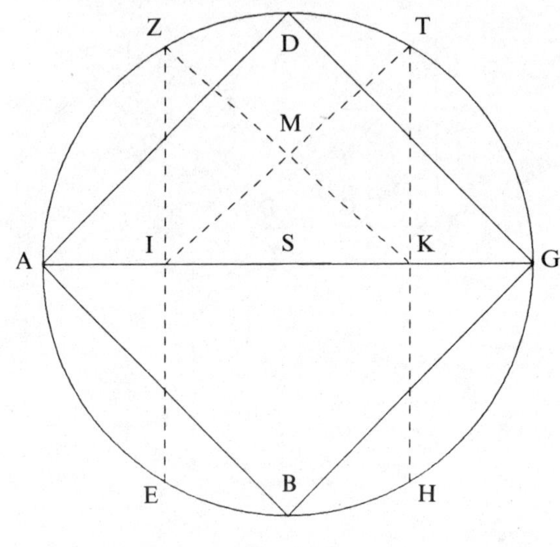

Fig. 3.19

Draw ZK *and* TI, *which intersect at* M, *and then draw the diameter through* S, M. *Let it meet the circle at* D *and* B. *Then* ADGB *will be a square.*

Proof. Since $\overset{\frown}{ZA} = \overset{\frown}{AE}$ the diameter GA bisects the arc $\overset{\frown}{ZE}$ and therefore GA is perpendicular to ZE, the chord of that arc. Similarly GA is perpendicular to TH, and so $\measuredangle$TKI and $\measuredangle$ZIK are right. Since TH and ZE are chords of equal arcs they are equal and therefore their halves, TK and ZI, are equal, and since they are also parallel (both being perpendicular to GA) the figure TKIZ is a rectangle. Its diagonals ZK and TI therefore are equal and bisect each other, and so MK = MI, i.e. $\triangle$(MKI) is isosceles. Since the equal chords ZE and TH are equidistant from the center (Euclid III,14), KS = SI, and so in the isosceles triangle MKI the line MS bisects the side KI and is therefore perpendicular to the side. Thus the diameter DB is perpendicular to the diameter GA and ADGB is a square.

As a final example of geometry with a rusty compass we take the following from Abu l-Wafā"s treatise.

Problem 5

To construct in a given circle a regular pentagon with a compass opening equal to the radius of the circle.

Procedure. *At the endpoint* A *of the radius* DA *erect* AE *perpendicular to* AD (Fig. 3.20), *and on* AE *mark off* AE = AD, *then bisect* AD *at Z and draw the line* ZE. *On this line mark off* ZH = AD *and bisect* ZH *at* T. *Then construct* TI *perpendicular to* EZ *and let* TI *meet* DA *extended at* I. *Finally, let the circle with center* I *and radius* AD *meet the given circle at points* M *and* L. *Then* $\overset{\frown}{ML}$

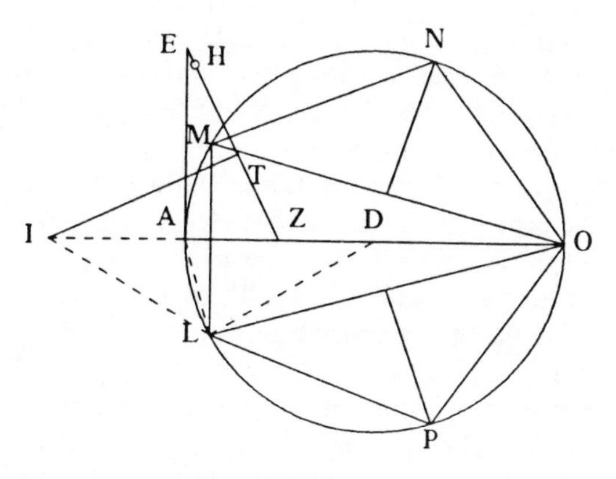

Fig. 3.20

is one-fifth of the circumference of the given circle, and the perpendicular bisector of the chord ML *bisects the complement of* $\widehat{\mathrm{ML}}$ *relative to the circle at* O. *Again, the perpendicular bisectors of the chords of* $\widehat{\mathrm{MO}}$ *and* $\widehat{\mathrm{LO}}$ *bisect the arcs themselves at* N *and* P, *respectively, and therefore the circle is divided into five equal arcs, whose chords will be the sides of the regular pentagon.*

Abu l-Wafā' proves the validity of the construction as follows. Draw the lines LD, LI and LA (Fig. 3.20). First of all, the triangles TIZ and AEZ are congruent since they are right triangles with the common angle Z and the side TZ = AZ. Thus EZ = ZI, and so:

$$\mathrm{ZI}^2 = \mathrm{EZ}^2 = \mathrm{EA}^2 + \mathrm{AZ}^2 = \mathrm{DA}^2 + \mathrm{AZ}^2.$$

Hence

$$\mathrm{DA}^2 = \mathrm{ZI}^2 - \mathrm{AZ}^2 = (\mathrm{ZI} + \mathrm{AZ})(\mathrm{ZI} - \mathrm{AZ}) = \mathrm{ID} \cdot \mathrm{IA}$$

and so A divides DI into two unequal parts so that the rectangle whose sides are the smaller part and the whole is equal in area to the square whose side is the larger part, the division into what the Greeks called "the section", although we now call it "the golden section". Now DA = LI, so $\mathrm{LI}^2 = \mathrm{ID} \cdot \mathrm{IA}$, which may be rewritten as the proportion ID/LI = LI/IA, and since in the two triangles LIA, DIL the angle I is common and the sides containing this angle are proportional it follows that the two triangles are similar. Thus DL/AL = LI/AI and, since DL = LI, AL = AI.

To finish the proof, we recall that Euclid proved in XIII,9 that if the side of a hexagon in a circle (AD) is extended in the direction of A by the side of a decagon in that circle then A divides the whole segment into the golden section, and the side of the hexagon is the larger of the two segments. Now, Abu l-Wafā' says, since the line segment ID has been divided in the golden section at A so that AD is the side of a hexagon in the circle (i.e. a radius) it follows that AI is the side of a decagon in the circle and hence the same holds for AL = AI. (Strictly speaking, Abu l-Wafā' needs to make a slight additional argument to deduce this from the converse of XIII,9, but it is a straightforward proof by contradiction and he left it out.) Thus AL is one-tenth of the circumference of the circle.

However, if two circles intersect at L and M then the line joining the two centers bisects the arcs between L and M. Hence $\widehat{LA} = \widehat{AM}$ and therefore $\widehat{LM}$ is one-fifth of the circumference of the given circle.

Abu l-Wafā''s treatise contains a wealth of beautiful constructions for regular n-gons, including exact constructions for $n = 3, 4, 5, 6, 8, 10$. It also gives a verging construction for $n = 9$ which goes back to Archimedes and the approximation for $n = 7$ that gives the side of a regular heptagon in a circle as equal to half the side of an inscribed equilateral triangle. This approximation, by no means original with Abu l-Wafā', was probably ancient even when Heron gave it in his *Metrica* in the first century A.D. However, it is a good approximation, and as a practical matter is much simpler than the exact construction by conics.

Exercises

1. Use the symptoms of the parabola and hyperbola given earlier to show in the "Third Reduction" in Section 4 that T lies on both the hyperbola and parabola.
2. In an Arabic manuscript found in Bankipore, the following construction of a regular nonagon inscribed in a circle is given: Let a circle with center D be quartered by two perpendicular diameters AE, ZH and let AB be a chord equal to a radius (Fig. 3.21). Let BTG be drawn so that it cuts the diameter ZH at T and the circle at G and so that TG = AB. Then TD is equal to the side of a regular nonagon inscribed in the circle ABGH. (1) Show that the point T is obtained by a verging construction. (2) Prove that TD is the side of a regular

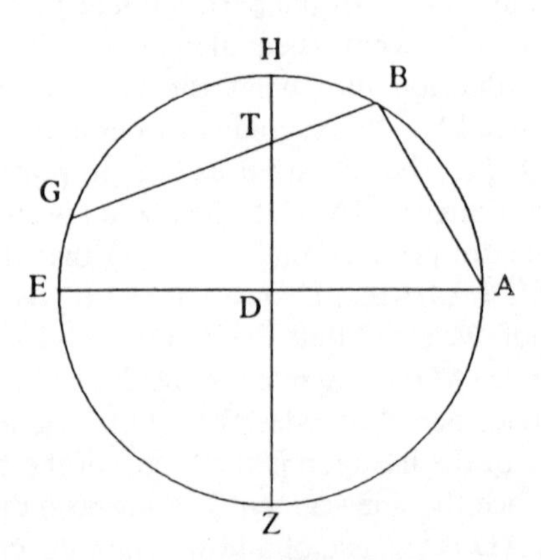

Fig. 3.21

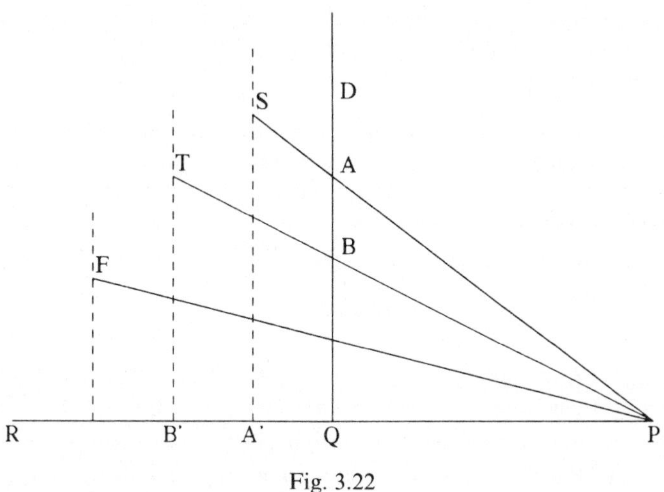

Fig. 3.22

nonagon in the circle. (*Hint*: If GL is perpendicular to DE and GM is perpendicular to DZ show that GL = DM = TM.)

3. The following method for constructing a segment of a parabola is found in a North African Arabic manuscript on burning mirrors now housed in the British Library. Let a line segment PR be given (Fig. 3.22), whose midpoint is Q, and at Q erect a segment QD perpendicular to PR. Divide DQ and RQ into the same number of equal segments. Suppose the points from D to Q are A, B, ... and the points from Q to R are A′, B′, ... At each of A′, B′, ... erect perpendiculars to RQ. Now let a straightedge passing through P and A intersect the perpendicular through A′ at S, let the straightedge through P and B intersect the perpendicular through B′ at T, etc. Then the points S, T, ... on the perpendiculars all lie on a parabola with vertex D. Using R in place of P, A″ in place of A′, etc. we obtain the other half of the parabola. Prove the validity of these statements.

4. In many textbooks on analytic geometry the student is asked to prove that if A and B are two given points in the plane then the set of all points X so that $|XA| - |XB| = k$, where k is a constant and $|XA|$, $|XB|$ are the distances of X from A and B, is a hyperbola. Use this fact to prove the validity of the following method, which Ibrāhīm gives for constructing a hyperbola. Let A be the center of a given circle and B be any point outside this circle. Consider the collection of all points X so that X is the center of a circle tangent to the given circle and passing through B. This collection is one branch of a hyperbola.

5. Show on the basis of Euclid XIII,9 that if AD is a radius of the circle and if $ID \cdot IA = AD^2$ then IA is the side of a regular decagon in the circle.

6. In a letter to Abu l-Jūd, al-Bīrūnī asks for a proof that Heron's construction for the side of a regular heptagon is not exact. Show this, but also show that Heron's construction would be in error by less than 2 mm in a circle of radius 1 m.

7. Show that the steps of Abū Sahl's analysis and the construction of the regular heptagon are equivalent, i.e. not only does the existence of $\triangle(ABG)$ imply the

division of a line segment such that (1) and (2) of the "Second Reduction" are satisfied but the converse is true as well.

8. Show that in Fig. 3.9 the exterior angle of $\triangle$(BAD) at B equals $3 \cdot \measuredangle$ D.
9. In the rusty-compass construction of a square in a given circle Abu l-Wafā' assumes his compass opening is equal to the radius of the circle. Show there is no loss of generality in this, that is that if the construction can be done in this special case then it can be done in the general case.

Bibliography

Berggren, J. L. "An Anonymous Treatise on the Regular Nonagon", *Journal for History of Arabic Science* **5** (1981), 37–41.

Daffa, A. A. and John Stroyls, "Naṣīr al-Dīn al-Ṭūsī's Attempt to Prove the Parallel Postulate of Euclid". In: *Studies in the Exact Sciences in Medieval Islam*. New York: Wiley, 1984.

Hogendijk, Jan P. "Greek and Arabic Constructions of the Regular Heptagon", *Archive for History of Exact Sciences* **30** (1984), 197–330.

Norman, Jane and Stef Stahl, *The Mathematics of Islamic Art: A Package for Teachers of Mathematics* ..., New York: Metropolitan Museum of Art, 1979.

Sabra, A. I. "Ibn al-Haytham's Lemmas for Solving 'Alhazen's Problem'", *Archive for History of Exact Sciences* **26** (1982), 299–324.

Winter, H. J. J. and W. 'Arafat, "Ibn al-Haitham on the Paraboloidal Focussing Mirror" and "A Discourse on the Concave Spherical Mirror of Ibn al-Haytham". *Journal of the Royal Asiatic Society of Bengal, Science* **15** (No. 1) (1949), 25–40; and **16** (No. 1) (1950), 1–16, respectively.

Woepcke, F. "Analyse et Extrait d'un Recueil de Constructions Géométriques par Aboûl Wafâ", *Journal Asiatique* (*Ser 5*), **5**, Feb.–March (1855), 218–359.
This contains an exposition, evidently based on a student's notes in Persian, of Abu l-Wafā''s fixed-compass constructions.

Item 2, 5 and 6 are included for students wishing to learn something of aspects of geometry in Islam not dealt with in this chapter.

Chapter 4

Algebra in Islam

§1. Problems About Unknown Quantities

Many ancient mathematical works contain problems requiring the discovery of an unknown quantity. Sometimes this is a geometrical magnitude which is related by the conditions of the problem to known magnitudes, one example being the problem solved in Euclid's *Elements* II,11 of dividing a given line segment AB into segments AG and GB so that the rectangle whose sides are AB and GB is equal to the square whose side is AG (Fig. 4.1). Here there is one known magnitude, AB, one unknown segment, AG, for $GB = AB - AG$, and the one condition that $AB \cdot GB = AG^2$. The reader will recall that this is the division of a line segment into "the section" we spoke of in Chapter 3 in our discussion of Problem 5 of Abu l-Wafā''s treatise in which a pentagon was inscribed in a circle.

Another example is found in Prop. 4 of Archimedes' *On the Sphere and Cylinder, Book II*, where Archimedes solves the problem of cutting a sphere by a plane so that the volumes of the two segments are to one another in a given ratio. In both of the above problems the unknown quantity is a geometrical magnitude, but examples where the unknown quantity is a number are abundant. For example, in a cuneiform text written in Mesopotamia, when the region was ruled by the successors of Alexander the Great, there is a problem asking for the number that, added to its reciprocal, produces a given number. Long before this, Babylonian scribes, as early as 1800 B.C., were able to solve problems that lead to quadratic equations, and, although the Babylonians would say "I have added the area and six times the side of my square and it makes 27" instead of "$x^2 + 6x = 27$" the procedure that they used for solving the problem is essentially the one that we use today.

The numerical procedures of the Babylonians are found again in the Greek writers Heron (fl. A.D. 60) and Diophantos (fl. A.D. 300), where sometimes more than one unknown is required. This is the case in the following problem solved by Diophantos in his *Arithmetica*: To find three

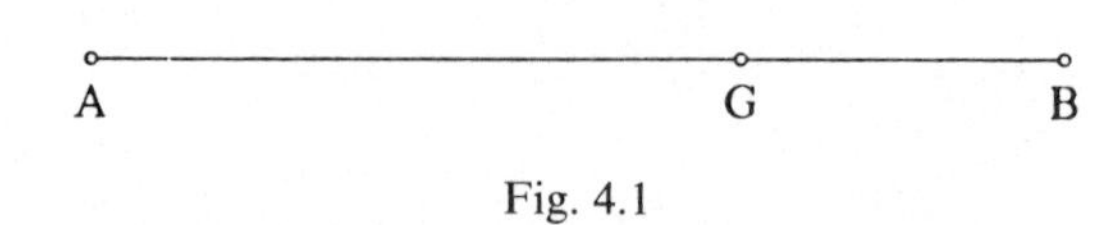

Fig. 4.1

(rational) numbers so that the product of any two added to the third gives a square. Other times it would be a single unknown subject to a variety of conditions, as in this problem found in the Hindu writer Bhaskara, around A.D. 1150: "Which number is it, which, being divided by six, has five for a remainder, and divided by five has a remainder of four, and divided by four has a remainder of three, and divided by three leaves two?".

For many of the Greek writers, however, the treatment of problems of finding unknowns was geometric. Thus, in his division of the line segment into "the section", Euclid is dealing with a geometric version of the problem of solving the quadratic equation $x^2 + ax = a^2$; for, if we let AB $= a$ and AG $= x$ (Fig. 4.1), then the condition

$$\mathrm{AB} \cdot \mathrm{GB} = \mathrm{AG}^2 \quad \text{becomes} \quad a(a - x) = x^2$$

and this implies $a^2 = ax + x^2$. This requires that, given a segment of length a, we construct a segment of length x so that x satisfies the condition $x^2 + ax = a^2$. Euclid, however, does not think of the numerical concept of "the length of a segment", but instead he looks on $x^2 + ax = x(x + a)$ as a rectangle having sides x and $x + a$. He literally "completes the square" when he uses Prop. II,6 of his *Elements* to the effect that if the square of side $a/2$ be added to this rectangle, $(x + a)x$, then the result is the square of side $x + a/2$. Then x is equal to the side of this square, $(x + a/2)$, less the known $a/2$.

The case of "the section" is but one example of how Greek geometers of the classic period saw such quadratic relationships as those we write as $x^2 + ax = a^2$ as statements about areas. To deal with these statements they had a body of theorems, e.g. those in Book II of the *Elements*, that allowed them to complete these areas to yield known squares, whose sides then gave the desired line segments. Many scholars believe that a historical reason for the Greek emphasis on geometry is that irrational numbers, such as $\sqrt{2}$ and $\sqrt{5}$, necessarily enter whenever one solves quadratic equations. Since the Greeks lacked even a definition of such numbers they could not treat them rigorously and so they used the geometrical magnitudes themselves when they wanted rigorous arguments.

§2. Sources of Islamic Algebra

All these parts of Greek mathematics were known to mathematicians of the Islamic world. The translation of Euclid done by al-Hajjāj late in the eighth century was followed by several improved translations, culminating in that of Thābit ibn Qurra late in the ninth century. Thābit also revised an earlier, imperfect translation of Archimedes' *Sphere and Cylinder*, so this work too was available in a good Arabic version from the tenth century onward. Finally, the first seven books of Diophantos' *Arithmetica* were translated into Arabic by Qusṭā ibn Lūqā of Baalbek, in present-day Lebanon, probably in the mid-ninth century. Both Qusṭā and Abu l-Wafā' wrote commentaries on this work.

Another great source of the Islamic algebra was the Hindu civilization, where from the late fifth century A.D. onwards we have abundant evidence of a highly developed mathematics. There are many parallels between the Hindu mathematicians and a Greek writer in the numerical tradition like Diophantos. Both used abbreviations of words to stand for unknowns; but, whereas Diophantos used only one abbreviation, the Hindu writers used many—the first syllables of the Sanskrit words for the various colors. Hindu writers had a special sign to denote negative numbers, and Diophantos, too, had a special sign for subtraction. Finally, both Diophantos and the Hindus were interested in indeterminate equations, that is, equations in several unknowns admitting a possibly infinite number of solutions.

One of the greatest early Hindu mathematicians was Brahmagupta, who lived in the first half of the seventh century A.D. and whose astronomical work, the *Brahmasphuta-siddhanta*, has two of its twenty-four chapters devoted to mathematics. In this work, Brahmagupta states clearly the rules for multiplying signed numbers and recognizes that the solutions to some of his problems may be negative numbers. He follows the earlier Āryabhaṭa (fl. A.D. 500) in giving the general solution to what we would write in the form $ax + by = c$ as $x = p + mb$, $y = q - ma$, where $x = p$, $y = q$ is a particular solution and m is any integer. He also showed how to use what we call "the Euclidean algorithm" to obtain particular solutions p and q. Thus $5x + 12y = 29$ has the solution $x = 1$, $y = 2$, and the general solution is $x = 1 + 12m$, $y = 2 - 5m$. He also discussed the difficult problem of solving $x^2 = 1 + py^2$, known today as Pell's equation.

The Muslims learned early in their history of the Hindu achievements in algebra, for Brahmagupta's astronomical work was one of those that Indian scholars brought to the caliph al-Manṣūr around A.D. 770, and it was translated by al-Fazārī into Arabic. Sanskrit astronomical works were written in verse form (perhaps to facilitate memorization), and the translation must have been no easy task for al-Fazārī. For example, al-Bīrūnī refers to the habit the early translators of Indian material had of leaving certain words untranslated, simply spelling them out in Arabic.

The Muslim mathematicians were as ready as their Babylonian and Hindu predecessors to appreciate the effectiveness of numerical procedures which these nations possessed for solving quadratic or indeterminate equations. As we have seen, they had inherited both the Babylonian sexagesimal system and the decimal system of the Hindus, and these efficient systems provided a good foundation for numerical mathematics.

On the other hand, the Greek geometrical approach had behind it the authority of men whom most Muslim scientists admired immensely, and the geometrical rules had been proved beyond any doubt to be true. Here it was not a case of checking a numerical answer to see that it solved the problem;

rather, there was a general theory based on sound axioms that yielded proofs of the validity of the methods.

In the next section we shall see how these two approaches, the numerical and the geometric, were combined to create a new science.

§3. Al-Khwārizmī's Algebra

The Name "Algebra"

Out of this dual heritage of solutions to problems asking for the discovery of numerical and geometrical unknowns Islamic civilization created and named a science—algebra. The word itself comes from the Arabic word "al-jabr", which appears in the title of many Arabic works as part of the phrase "al-jabr wa al-muqābala". One meaning of "al-jabr" is "setting back in its place" or "restoring", and the ninth century algebraist al-Khwārizmī, although he is not always consistent, uses the term to denote the operation of restoring a quantity subtracted from one side of the equation to the other side to make it positive. Thus replacing $5x + 1 = 2 - 3x$ by $8x + 1 = 2$ would be an instance of "al-jabr". The word "wa" just means "and", and it joins "al-jabr" with the word "al-muqābala", which means in this context replacing two terms of the same type, but on different sides of an equation, by their difference on the side of the larger. Thus, replacing $8x + 1 = 2$ by $8x = 1$ would be an instance of "al-muqābala".

Clearly, with the two operations any algebraic equation can be reduced to one in which a sum of positive terms on one side is equal either to a sum of positive terms involving different powers of x on the other, or to zero. In particular, any quadratic equation with a positive root can be reduced to one of three standard forms:

$$px^2 = qx + r, \quad px^2 + r = qx, \quad \text{or} \quad px^2 + qx = r, \quad \text{with } p, q, r \text{ all positive,}$$

a condition that runs through the whole medieval period in Islamic mathematics. We shall meet it again in the work of ʿUmar al-Khayyāmī, and it is the rule in Western mathematics as well through the early sixteenth century. Thus the science of "al-jabr wa al-muqābala" was, at its beginning, the science of transforming equations involving one or more unknowns into one of the above standard forms and then solving this form.

Basic Ideas in Al-Khwārizmī's Algebra

One of the earliest writers on algebra was Muḥammad b. Mūsā al-Khwārizmī, whose treatise on Hindu reckoning we referred to in Chapter 2. His work on algebra, *The Condensed Book on the Calculation of al-Jabr wa al-Muqābala*, enjoyed wide circulation not only in the Islamic world but in the Latin West as well.

According to al-Khwārizmī there are three kinds of quantities: *simple numbers* like 2, 13 and 101, then *root*, which is the unknown, x, that is to be found in a particular problem, and *wealth*, the square of the root, called in Arabic *māl*. (A possible advantage of thinking of the square term as representing wealth is that al-Khwārizmī can then interpret the number term as *dirhams*, a local unit of currency. Another word used for "root" by many writers is "thing". In these terms al-Khwārizmī could list the six basic types of equations as:

(1) Roots equal numbers ($nx = m$).
(2) Māl equal roots ($x^2 = nx$).
(3) Māl equal numbers ($x^2 = m$).
(4) Numbers and māl equal roots ($m + x^2 = nx$).
(5) Numbers equal roots and māl ($m = nx + x^2$).
(6) Māl equals numbers and roots ($x^2 = m + nx$).

All equations involving only the three basic quantities and having a positive solution could be reduced to one of these three types, the only ones with which al-Khwārizmī concerns himself.

Al-Khwārizmī's Discussion of $x^2 + 21 = 10x$

In following al-Khwārizmī's discussion of type (4) above we shall use modern notation to render his verbal account. He discusses this type in terms of the specific example $x^2 + 21 = 10x$, which he describes as "māl and 21 equals 10 roots", as follows (translation adapted from F. Rosen):

> Halve the number of roots. It is 5. Multiply this by itself and the product is 25. Subtract from this the 21 added to the square (term) and the remainder is 4. Extract its square root, 2, and subtract this from half the number of roots, 5. There remains 3. This is the root you wanted, whose square is 9. Alternately, you may add the square root to half the number of roots and the sum is 7. This is (then) the root you wanted and the square is 49.

Notice that al-Khwārizmī's first procedure is simply a verbal description of our rule

$$\frac{10}{2} - \sqrt{\left(\frac{10}{2}\right)^2 - 21},$$

and his second procedure describes the calculation of $5 + \sqrt{5^2 - 21}$, but since all quantities are named in terms of their role in the problem whenever they appear (For example, "5" is called "the number of roots"), his description of the solution is quite as general, if not so compact, as our

$$\frac{n}{2} \pm \sqrt{\left(\frac{n}{2}\right)^2 - m}.$$

In fact, al-Khwārizmī's generality is reflected in the remarks that continue

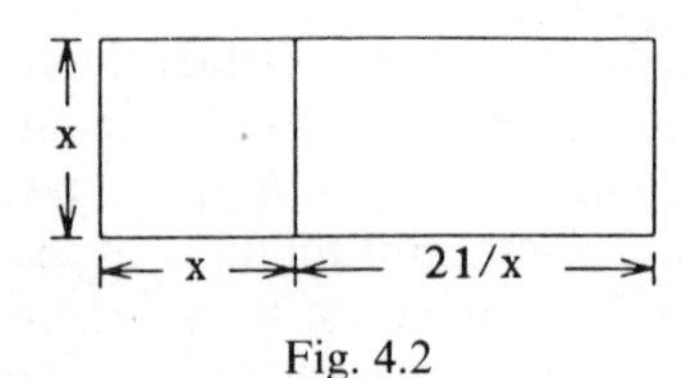

Fig. 4.2

those quoted above:

> When you meet an instance which refers you to this case, try its solution by addition, and if that does not work subtraction will. In this case, both addition and subtraction can be used, which will not serve in any other of the three cases where the number of roots is to be halved.
>
> Know also that when, in a problem leading to this case, you have multiplied half the number of roots by itself, if the product is less than the number of dirhams added to māl, then the case is impossible. On the other hand, if the product is equal to the *dirhams* themselves, then the root is half the number of roots.

In the first of the above paragraphs al-Khwārizmī recognizes that the case we are dealing with is the only one where there can be two positive roots. In the second paragraph he remarks that there is no solution when what we call the discriminant is less than zero and he says that when $(n/2)^2 = m$ the only solution is $n/2$. Finally, he remarks that in the case $px^2 + m = nx$ it is necessary to divide everything by p to obtain $x^2 + (m/p) = (n/p)x$, which can be solved by the previous method. This shows, by the way, that his coefficients are not restricted to whole numbers.

What distinguishes al-Khwārizmī and his successors from earlier writers on problems of the above sort is that, following the procedures for obtaining the numerical solutions, he gives proofs of the validity of these same procedures, proofs that interpret $x^2 + 21$, for example, as a rectangle consisting of a square (x^2) joined to a rectangle of sides x and $21/x$ (Fig. 4.2).

§4. Thābit's Demonstration for Quadratic Equations

Preliminaries

Al-Khwārizmī presents his proofs in terms of particular equations, but Thābit ibn Qurra in his work gives the demonstrations in general, and for that reason we shall follow him rather than the earlier al-Khwārizmī.

The first two cases, $x^2 + px = q$ and $x^2 + q = px$ sufficiently indicate Thābit's approach. In the proofs he uses two theorems from Euclid's *Elements*, which we now state and prove.

Book II, Prop. 5. *If a line* AE *is divided at* B *and bisected at* W *then the*

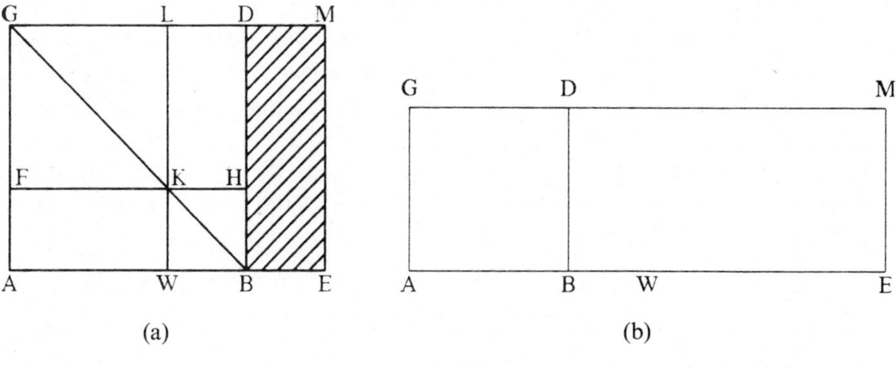

Fig. 4.3

rectangle AB · BE *plus the square on* BW *is equal to the square on* AW (Fig. 4.3(a) *and* (b)).

Note that in this proposition B may be on either side of the midpoint W. The two parts of Fig. 4.3 show these two cases and are drawn so that GAEM is a rectangle of sides AE and AG(= AB). The rectangle AB · BE which the theorem speaks of is equal to the shaded rectangle since AB = BD. The foregoing proposition deals with a line segment bisected and divided internally. The next proposition deals with a line segment extended, which we could look on as being bisected and divided externally.

Book II, Prop. 6. *If a given line* BH *is bisected at* W *and extended in a straight line* BA *then rectangle* AH · AB *plus the square on* BW *is equal to the square on* AW (Fig. 4.4).

Proof of II,5. In the case where B is between W and E the truth of the theorem is most clearly seen by drawing the diagonal GB in the square

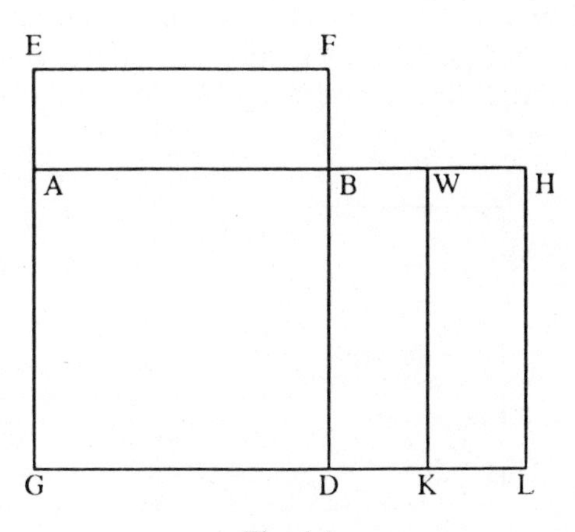

Fig. 4.4

ABDG and the line WKL parallel to BD. Finally, draw the line FKH parallel to AB. ABDG is a square since AG = AB, so WBHK is also a square. Now consider the L-shaped figure WEMDHK. Such a figure was called a "gnomon" by the Greeks. In the case of the gnomon WEMDHK, which is made up of the rectangle AB · BE plus the square on BW, it is what remains when KHDL is taken away from the rectangle WEML. Also KHDL = AWKF, by Euclid I,43, while WEML = WAGL, since AW = WE. Thus the gnomon

$$\text{WEMDHK} = \text{WEML} - \text{KHDL} = \text{WAGL} - \text{AWKF} = \text{FKLG},$$

and FKLG is equal to the square on AW since it is a square and its side FK = AW.

The proof for the other case of this theorem is similar and is left as an exercise.

Proof of II,6. Construct the square on AB, say ABDG, and the rectangle DBHL. Then GAHL equals the rectangle AH · AB since AB = AG. The parallel to DB through W divides DBHL into two congruent rectangles, since BW = WH. If WHLK is placed above AB so that WK coincides with AB then we see that the original rectangle with sides AH, AB is equal to the gnomon EGKWBF; but, if the square BF^2 is added to this figure, the result is the square on GK. The proof is now complete since the square on GK is the square on AW, because AW = GK.

Thābit's Demonstration

Thābit begins his discussion of the validity of the procedures for solving quadratic equations by discussing the solution of what he calls the first basic form, "māl and roots equal numbers." In following the discussion the reader should know that Thābit's term "roots" corresponds to the modern "px" and that he uses the words "the number of roots" to refer to the coefficient "p". In Fig. 4.5 let the square ABDG represent māl (so its side AB is a root), and suppose the unit of linear measure is chosen so that BH represents the

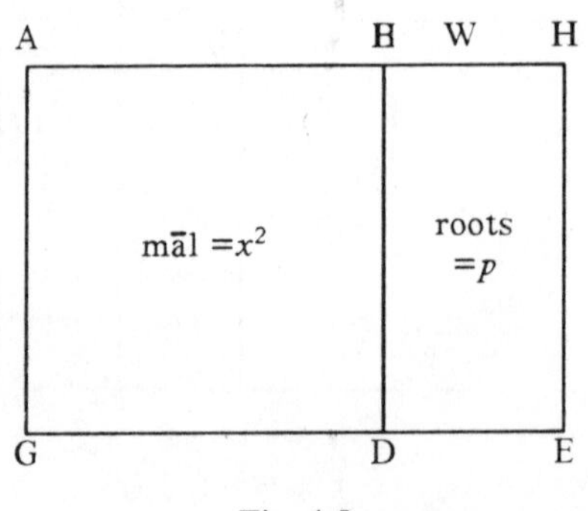

Fig. 4.5

number of roots. The area DEHB then represents the term "roots", and so GEHA represents māl and roots. According to Euclid II,6, then, GEHA plus BW^2 is equal to AW^2. But māl and roots is known (for it is "numbers"), and BW^2 is also known since BW is half the given number of roots. Hence AW^2, and thus AW, is known. But $x = AB = AW - BW$, so x is known.

Finally, Thābit shows the correspondences between the geometric solution given and the algebraic solution.

$$\tfrac{1}{2}\mathrm{BH} = \mathrm{BW} \leftrightarrow \text{half the number of roots,}$$

$$\text{Square on BW} \leftrightarrow \text{square of the above,}$$

$$\text{Rectangle on } \mathrm{HA} \cdot \mathrm{AB} \leftrightarrow \text{numbers,}$$

$$\text{Square on AW} \leftrightarrow \text{sum of the two preceding,}$$

$$\mathrm{AW} \leftrightarrow \text{square root of the sum,}$$

$$\mathrm{AW} - \mathrm{BW} = \mathrm{AB} \leftrightarrow \text{the square root less half the number of roots.}$$

He adds $(p/2)^2$ to each side to obtain

$$x^2 + px + \left(\frac{p}{2}\right)^2 = q + \left(\frac{p}{2}\right)^2.$$

With his geometrical interpretation of all terms as areas he is able to apply Euclid II,6 to the left-hand side to obtain

$$\left(x + \frac{p}{2}\right)^2 = q + \left(\frac{p}{2}\right)^2$$

and therefore x is determined by

$$x = \sqrt{q + \left(\frac{p}{2}\right)^2} - \frac{p}{2}.$$

Thābit then shows the validity of the solution for the second basic form, "māl plus numbers equal to roots", as follows (Fig. 4.6(a), (b)). Again, the square ABDG represents māl and on AB extended the point E is chosen so that, a unit of measure being given, the length of the line AE is equal to the number of roots. (Thābit in fact shows that such a point E must lie on the extension of AB.) Now, bisect AE at W and construct the rectangle with sides GA and AE. Notice that GA is "root" and AE is the number of roots, so the foregoing rectangle represents "roots", which is "māl plus numbers", and the square (BG) is māl. (We use two opposite corners to denote a quadrilateral.) Thus when we subtract māl the remaining rectangle is "numbers". According to Euclid II,5, "numbers" plus the square on BW is equal to the square on AW. But this square is known, since AW is half the number of roots, and "numbers" is also known, so the remaining term, the square on BW, is known, and hence BW itself is known. Since AW and WB are both known, AB (which is "root") is also known—in the first diagram as $AB = AW + WB$, and in the second as $AW - WB$.

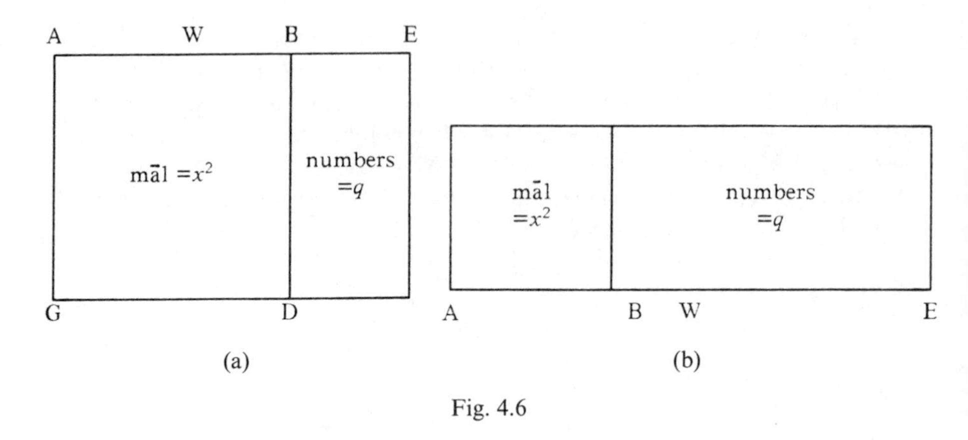

Fig. 4.6

Thābit does not go into details on the algebra, since he evidently feels the correspondences between the algebra and the geometry are sufficiently clear from the previous case.

§5. Abū Kāmil on Algebra

Similarities with al-Khwārizmī

A writer who was active about the time of Thābit's death in 901 is Abū Kāmil, whose epithet "the Egyptian calculator" tells us practically all we know about him. His work *Algebra* was a commentary on that of al-Khwārizmī and, perhaps in part for that reason and in part for its own merits, the book became quite popular. Both the Muslim writer al-Karajī in the late tenth century and the Italian Leonardo of Pisa, known as Fibonacci, in the late twelfth century made considerable use of Abū Kāmil's examples.

As one would expect with a commentary there are many similarities between Abū Kāmil and al-Khwārizmī. For example, he follows al-Khwārizmī in naming the basic quantities numbers, roots and māl. Like al-Khwārizmī's work, the *Algebra* is entirely verbal, and even the numbers are written out. In addition there is the same classification of equations, with the six types appearing in the same order and, for the equations involving all three terms, there are the same examples. Finally, like al-Khwārizmī, Abū Kāmil discusses the geometrical proofs of the procedures in terms of specific examples rather than in general as Thābit does.

Advances Beyond al-Khwārizmī

Despite this Abū Kāmil's work goes beyond that of al-Khwārizmī in giving general statements of rules which al-Khwārizmī states by means of examples. In addition he provides proofs of such rules for manipulating algebraic quantities as the following:

(1) $(a \pm px)(b \pm qx) = ab \pm bpx \pm aqx + pqx^2$,
$(a \pm px)(b \mp qx) = ab \pm bpx \mp aqx - pqx^2$.

(2) $\sqrt{a \cdot b} = \sqrt{a} \cdot \sqrt{b}$ and $\sqrt{a/b} = \sqrt{a}/\sqrt{b}$.

(3) $\sqrt{a} \pm \sqrt{b} = \sqrt{a + b \pm 2\sqrt{ab}}$.

The identities in (1) are obviously fundamental, while (2) and (3) taken together allow one to write arithmetic combinations of square roots as square roots. In addition, (2) is of some practical utility in that it allows us to calculate, say, $\sqrt{13} \cdot \sqrt{5}$ by multiplying only whole numbers and taking only one square root, namely $\sqrt{13 \cdot 5} = \sqrt{65}$, which is now easy to see as a bit over 8, something not clear from $\sqrt{13} \cdot \sqrt{5}$.

In the case of (1), Abū Kāmil expresses the rule as, "When roots are added to numbers or subtracted from them, on whichever side they are arranged (i.e. $a + px$ or $+ px + a$), then the fourth part is added, where "the fourth part" is the product of the roots, one by the other." Here he speaks only of "the fourth part", i.e. $(+px) \cdot (+qx)$ or $(-px) \cdot (-qx)$ since he assumes his readers know the signs of the first three terms of these products. A little later he summarizes the rules of signs for multiplication, namely $(-a)(-b) = +(ab)$, $(-a)(+b) = -(ab)$, and $(+a)(+b) = +(ab)$, as, "The product, when the two terms are subtracted, is added; the subtracted times the added is subtracted; the added times the added is added."

Abū Kāmil carefully demonstrates rules like $ax \cdot bx = ab \cdot x^2$ and $a \cdot (bx) = (ab) \cdot x$ (where a, b are always specific numbers in the demonstrations), and then he shows the truth of particular cases of (1), e.g.

$$(10 - x) \cdot (10 - x) = 100 + x^2 - 20x.$$

Although Abū Kāmil gives an algebraic proof, based on the distributive law and the rule of signs, he also gives the following geometric proof.

Proof (Fig. 4.7). Let the line GA represent the number 10 and GB the "thing", x, and complete the square (AD) as in the diagram. Then AB =

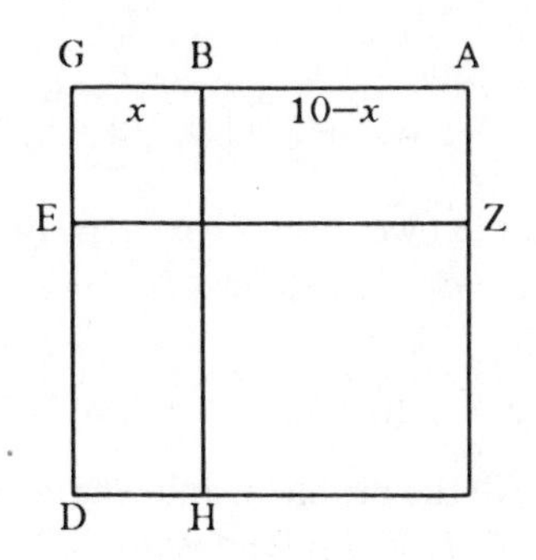

Fig. 4.7

ED $= 10 - x$, so the square (ZH) $= (10 - x)^2$. Also, (GZ) = (GH) $= 10x$, so (EH) = (GH) − (EB) $= 10x - x^2$. Hence, the gnomon (EH) + (GZ) $= 20x - x^2$. Since the large square is 100 it follows that

$$(10 - x)^2 = (\mathrm{ZH}) = 100 - (20x - x^2) = 100 + x^2 - 20x.$$

Abū Kāmil gives no explanation for the last equality, so he presumably feels it is clear either from the rule of signs or from the figure.

The samples from Thābit and Abū Kāmil suffice to illustrate the demonstrations algebraists of the Islamic world gave for algebraic rules by geometric theorems, and we omit the demonstrations of (2) and (3).

A Problem from Abū Kāmil

Abū Kāmil's work includes a great variety of problems, 69 in all, amounting to almost 30 more than the 40 problems al-Khwārizmī explained. One of the most interesting is Problem 61, which Abū Kāmil states as follows:

> One says that 10 is divided into three parts and if the small one multiplied by itself is added to the middle one multiplied by itself the result is the large one multiplied by itself, and when the small is multiplied by the large it equals the middle multiplied by itself.

Abū Kāmil thus speaks of three unknown quantities x, y, z (assumed to be positive) satisfying the three conditions

$$10 = x + y + z, \qquad z^2 = x^2 + y^2 \quad \text{and} \quad xz = y^2.$$

He first sets $x = 1$ and the conditions become

$$10 = 1 + y + z, \qquad z^2 = 1 + y^2 \quad \text{and} \quad z = y^2.$$

The last two yield $(y^2)^2 = 1 + y^2$, and this quadratic in y^2 he solves to obtain

$$z = y^2 = \tfrac{1}{2} + \sqrt{1\tfrac{1}{4}}, \quad \text{so} \quad y = \sqrt{\tfrac{1}{2} + \sqrt{1\tfrac{1}{4}}}.$$

Then

$$1 + y + z = \tfrac{3}{2} + \sqrt{1\tfrac{1}{4}} + \sqrt{\tfrac{1}{2} + \sqrt{1\tfrac{1}{4}}}.$$

If we call this quantity a then we should have $a = 10$, but it is obviously not equal to 10; however, $a \cdot 10/a = 10$, so if we set $10/a = b$ then $a \cdot b = 10$. This may be written

$$(1 + y + z) \cdot b = 10, \quad \text{i.e.} \quad b + (yb) + (zb) = 10.$$

Hence b, yb and zb solve the problem, and what Abū Kāmil has used here is something known as "the rule of false position". This algebraic device is, if not as old as the hills, at least as old as the pyramids, since it is found in ancient Egyptian texts. In its simplest form one would solve $5x = 24$ by saying, "When $x = 1$, $5 \cdot x = 5$. What must I multiply 5 by to make 24? The answer is $\frac{24}{5} = 4\frac{4}{5}$. Thus the true value of x is $1(4\frac{4}{5})$." Of course in Abū Kāmil's problem there are more variables and more conditions, but in each condition (equation) every term is of the same degree (one or two) and therefore the rule of false position holds valid.

We shall not follow Abū Kāmil's calculation of the value of b, and shall only remark that it is a virtuoso performance with the rules of algebra, including his use, without comment, of the identity

$$\sqrt{11\tfrac{1}{4} \cdot x^2 \cdot x^2} - \sqrt{1\tfrac{1}{4} \cdot x^2 \cdot x^2} = \sqrt{5 \cdot x^2 x^2}.$$

In the end he obtains b as a root of the equation:

$$10x = x^2 + 75 - \sqrt{3125}.$$

This equation is one of the six standard forms, and he solves it to obtain:

$$b = 5 - \sqrt{\sqrt{3125} - 50}.$$

A similar procedure leads him to an expression for z, the largest root, as

$$2\tfrac{1}{2} + \sqrt{31\tfrac{1}{4}} - \sqrt{\sqrt{78\tfrac{1}{4}} - 12\tfrac{1}{2}}$$

and from these two, knowing that $x + y + z = 10$, he is able to write down an expression for y.

The solution to this problem, whose key steps are the use of the rule of false position and the identity for $\sqrt{a} - \sqrt{b}$ shows that authors in the tenth century were capable of rationalizing denominators in expressions of the form $a/(c + \sqrt{d} + \sqrt{e} + \sqrt{f})$, which involved dealing with powers of the unknown as high as the eighth and solving quadratic equations with irrational numbers as coefficients.

§6. Al-Karajī's Arithmetization of Algebra

Introduction

Abū Kāmil's work shows the development of an arithmetic of expressions of the form $a + b\sqrt{q}$, where a, b and q are rational numbers with q nonnegative. Although Thābit and Abū Kāmil apply geometry to algebra, there is in Abū Kāmil's treatment of the numbers $a + b\sqrt{q}$ a tendency to apply arithmetic to a subject that, long before him, Euclid had treated geometrically. We have also seen increasingly complicated examples forcing algebraists to consider more complicated expressions than those involving only "things" or "squares", and both of these tendencies are carried further in the algebra named *The Marvellous*, composed by Abū Bakr al-Karajī.

Al-Karajī is one of the many remarkable Muslim scientists about whose life we should like to know more than we do, but we do know that he worked in Baghdad around the year A.D. 1000 and that in the first decade of that century he dedicated a book on algebra to a vizier Fakhr al-Mulk. Sometime later he left for "the mountain countries". He must have been a man of wide interests, for among his writings appear not only treatises on arithmetic, algebra, indeterminate analysis and astronomy, but on surveying and on finding underground waters as well.

Although writers like Diophantos and Abu l-Wafā' indicated the possibility of using arbitrarily high powers of the unknown, al-Karajī appears to have been the first person to develop the algebra of expressions containing these powers. His point of view is that unknown quantities, whether absolute numbers or geometrical magnitudes, can be a "root", "side" or "thing" (both corresponding to our "x"), or they can be māl (x^2), cube (x^3), māl māl (x^4), māl cube (x^5), etc., where each member is the product of "thing" by the previous member. These different species of quantity al-Karajī calls "orders" (a terminology also used for the places of the different powers of 10 in decimal arithmetic), and he observes that the number common to all orders is 1 (since it is equal to all its powers). In addition, corresponding to each order (x^n) is the corresponding part ($1/x^n$), with the property that any order times its part equals 1. On this basis al-Karajī developed his program of treating expressions like "māl māl and four cubes less six units" ($x^4 + 4x^3 - 6$) and "five cube cubes less two squares and three units" ($5x^6 - [2x^2 + 3]$) by rules modelled on the ordinary rules of arithmetic for adding, subtracting, multiplying, dividing and extracting square roots.

R. Rashed has referred to this modelling of the algebra of polynomials on positional arithmetic as the "arithmetization of algebra". Al-Karajī was one of the pioneers in this process, and if his success in arithmetizing algebra was only partial it is due less to lack of ingenuity than to the lack of a way of incorporating negative numbers into the theory. Thus, although al-Karajī knew rules like $a - (-b) = a + b$, for a and b positive, he evidently had not discovered the rule $-a - (-b) = -(a - b)$. This prevented him from extending his method for dividing two polynomials to cover all cases, for his procedure, which we shall explain later, would, in general, require the subtraction of one negative quantity from another. It also prevented his discovery of a way to extract square roots of polynomials, for the same reason. Students who have struggled with the law of signs may find comfort in learning that at one time the discovery of these rules taxed the ingenuity of the best mathematicians, and that the discovery of much of our elementary (pre-calculus) mathematics was a matter of considerable labor and many false starts.

We do, however, find the laws for dealing with signed magnitudes in the writings of a physician named al-Samaw'al ben Yaḥyā ben Yahūdā al-Maghribī, who was born in Baghdad, perhaps 70 years after al-Karajī died, and whose work is a commentary on that of al-Karajī. In his work *Al-Bāhir fi'l-Ḥisāb* (*The Shining Book on Calculation*), which he wrote when he was nineteen, al-Samaw'al states the more troublesome parts of the rule of signs as follows:

> ... if we subtract a deficient (negative) number from a deficient number larger than it (i.e. representing a greater deficiency) there remains the difference, deficient (e.g. $-5-(-2)=-(5-2)$), but in the other case there remains their difference, excess ($-2-(-5)=+(5-2)$). If we subtract an excess number from an empty order (one where there is a zero), the same number, deficient, remains; but, if we subtract a deficient number from an empty order there remains in it that number, excess.

Thus, al-Samaw'al conceived of numbers as expressing either an excess (our "positive") or a deficiency ("negative"), so that, for example, a number expressing a deficiency of 5 would be larger than one expressing a deficiency of 2. Al-Samaw'al's rules for subtracting powers, when expressed symbolically, say that $(-ax^n)-(-bx^n)$ is equal to

$$-(ax^n - bx^n), \qquad \text{if} \quad a > b$$

and to

$$+(bx^n - ax^n), \qquad \text{if} \quad a < b.$$

With these rules al-Samaw'al could, and did, use the procedures we know today for adding and subtracting combinations of powers by adding and subtracting like terms.

Al-Samaw'al was born into a Jewish family, and it is a commentary on the conditions in Baghdad at that time that he could find no one competent to teach him mathematics beyond the first few books of Euclid's *Elements*. So he completed his studies of that work by himself and then studied the works of Abū Kāmil and al-Karajī. In his autobiography he tells of a dream he had in 1163 that made him convert to Islam. His life was spent as a travelling medical doctor, who counted princes among his patients, and he died in Maragha—in northern Iran—around 1180.

Of Al-Samaw'al's eighty-five recorded works, which range over mathematics and astronomy, medicine and theology, only a few have survived, and, of the surviving mathematical works, we shall discuss only *The Shining*, which we have mentioned earlier. Our account will deal with two sections from the first part on algebra, namely those dealing with the law of exponents and dividing expressions composed of different orders.

Al-Samaw'al on the Law of Exponents

The basis for al-Samaw'al's rules for multiplication is the following chart. In presenting this and other pieces from al-Samaw'al's work we use, as a compromise between the modern and medieval notation, abbreviations, such as "mcc" for "māl cube cube" (x^8) and "pcc" for "part of cube cube" ($1/x^6$).

I pccc	H pmcc	G pmmc	F pcc	:E pmc	D pmm	C pc	B pm	A pt	0 unit	A t	B m	C c	D mm	E mc	F cc	G mmc	H mcc	I ccc
$\frac{1}{8}\frac{1}{8}\frac{1}{8}$	$\frac{1}{4}\frac{1}{8}\frac{1}{8}$	$\frac{1}{4}\frac{1}{4}\frac{1}{8}$	$\frac{1}{8}\frac{1}{8}$	$\frac{1}{4}\frac{1}{8}$	$\frac{1}{4}\frac{1}{4}$	$\frac{1}{8}$	$\frac{1}{4}$	$\frac{1}{2}$	1	2	4	8	16	32	64	128	256	512
$\frac{1}{27}\frac{1}{27}\frac{1}{27}$	$\frac{1}{9}\frac{1}{27}\frac{1}{27}$	$\frac{1}{9}\frac{1}{9}\frac{1}{27}$	$\frac{1}{27}\frac{1}{27}$	$\frac{1}{9}\frac{1}{27}$	$\frac{1}{9}\frac{1}{9}$	$\frac{1}{27}$	$\frac{1}{9}$	$\frac{1}{3}$	1	3	9	27	81	243	729	2,187	6,561	19,683

Chart 4.1

In this way the reader may gain a better feeling for mathematics in the twelfth-century Muslim world.

To begin, al-Samaw'al sets down a chart aimed at teaching the reader how to multiply or divide simple expressions, such as pmc and mmc. The columns of this chart are headed by the usual Arabic alphabetic numerals, so A stands for 1, B for 2, etc. (Chart 4.1). The fourth column to the left of the one headed "0", namely the one headed "D", is read: "When 'thing' is 2 (resp. 3) then 'part of māl māl' is ($\frac{1}{4}$) ($\frac{1}{4}$) (resp. ($\frac{1}{9}$)($\frac{1}{9}$))."

The importance of the chart is that al-Samaw'al uses it to obtain an operational equivalent of the law of exponents: For any two integers m and n, $x^m x^n = x^{(m+n)}$. He states the rule as follows, where "distance" means the "number of cells on the chart":

> The distance of the order of the product of the two factors from the order of one of the two factors is equal to the distance of the order of the other factor from the unit. If the factors are in different directions then we count (the distance) from the order of the first factor towards the unit; but, if they are in the same direction, we count away from the unit.

Al-Samaw'al discusses several examples of this rule in which he makes good use of the numerals above the orders. Thus in the example of pc times pmm he says that since mm is fifth, starting from the unit, we count, starting at pc and moving away from the unit, five orders and end up at pmmc; but, then he says:

> Opposite (above) the order of part of cube is 3 and opposite part of māl māl is 4. We add them to obtain 7 and opposite (below) it is the order of part of māl māl cube.

Thus it was that a twelfth-century algebraist expressed what we write as $x^{-3}x^{-4} = x^{-7}$. Again, al-Samaw'al writes (where we substitute numerals for his number-words):

> To find 3 parts of māl multiplied by 7 cubes we multiply 3 by 7 and 21 results. We find the order of "cube" to be fourth from the unit, so we count four orders from the order of "part of māl" towards the unit, so that the result ... is 21 things.... If we want, we take the difference of the numbers opposite the orders of the factors, namely 2 and 3, and we find 1, and opposite 1, we find, in the direction of the factor that has the larger number opposite it, the order of things.

This latter rule we may write as $x^n x^{-m} = x^{(n-m)}$, for n greater than m.

Al-Samaw'al's justification for his general rule is that if $c = a \cdot b$ then $c:a = b:1$, so that the product c is related to a as b is related to 1. In particular, if b is n orders to the right or left of the order of 1 then c must be n orders to the right or left of the order of a. This is a nice example of giving a paraphrase of a rigorously defined mathematical relation and using the paraphrase, interpreted the way one feels is right, to deduce consequences of the relation. In the hands of one with a sure feeling for the subject such heuristic reasoning is a fruitful method of discovery, and so it was with al-Samaw'al.

With the above rules for multiplying the individual orders, al-Samaw'al has no difficulty explaining how to multiply two expressions, each composed of various orders, simply by multiplying each term of one by all terms of the other and adding up the results.

As for division, he notes that, "The division of a composite expression by a single order is easy for him who knows the division of the single (by the single), and the division of the single is easy for him who knows the multiplication of the single." (The first, of course, is because $(a + b)/c = a/c + b/c$, and the second because $a/c = d$ means $a = d \cdot c$. He says, however, that other cases are more difficult, and that no one up to his time has solved them, but he has found a way and gives the following example to clarify it.

Al-Samaw'al on the Division of Polynomials

The First Example

He sets the problem of dividing:

$$20\,\text{cc} + 2\,\text{mc} + 58\,\text{mm} + 75\,\text{c} + 125\,\text{m} + 196\,\text{t} + 94\,\text{units} + 40\,\text{pt}$$
$$+\ 50\,\text{pm} + 90\,\text{pc} + 20\,\text{pmm}$$

by 2 c + 5 t + 5 units + 10 pt. In modern symbols this asks for the computation of

$$\frac{20x^6 + 2x^5 + 58x^4 + 75x^3 + 125x^2 + 196x + 94 + 40x^{-1} + 50x^{-2} + 90x^{-3} + 20x^{-4}}{2x^3 + 5x + 5 + 10x^{-1}}$$

He says that to start, "We put the two terms in natural arrangement, and in each empty order we put a zero." (Al-Samaw'al's procedure is obviously intended to be used on the dust board, where erasure is easy, but space is at a premium, and it proceeds by a series of charts. It adapts easily, however, to paper, where erasure is not easy but space is ample, and nothing is lost if we combine his charts into one (Chart 4.2).

The top row lists the names of the orders in their natural sequence from left to right, and the row below it is the row of the answer, which is initially empty and is filled in as the process proceeds. The rest of the chart is divided

cc	mc	mm	c	m	t	unit	pt	pm	pc	pmm
			10	1	4	10	0	8	2	
20	2	58	75	125	96	94	140	50	90	20
2	0	5	5	0	0	0	10			
	2	8	25	25	96	94	140	50	90	20
	2	0	5	5	10					
		8	20	20	86	94	140	50	90	20
		2	0	5	5	10				
			20	0	66	54	140	50	90	20
			2	0	5	5	10			
					16	4	40	50	90	20
					2	0	5	5	10	
						4	0	10	10	20
						2	0	5	5	10

Chart 4.2

into horizontal bands, each containing two rows, and each band, together with the two top rows, constitutes one of al-Samaw'al's charts. Thus we may speak of the first, second, etc. charts, and the reader will understand what we mean.

In the first chart are the coefficients of the dividend, each under the name of its order, and underneath, starting in the left-most column, are the coefficients of the divisor. Notice that whereas the names above the terms of the dividend correspond to the names of the corresponding orders this is not the case with the divisor, so you must remember that the divisor begins, in this case, with cubes. Now, says al-Samaw'al,

> We divide the greatest order of the dividend (20 cc) by the greatest order of the divisor (2 c) and the result, 10 cubes, we place in the order facing 75 (the order of cubes). We next multiply it (10 c) by the divisor, and we subtract the result of its multiplication by each order (of the dividend) from what is above it.

That is, al-Samaw'al divides 20 cc by 2 c to obtain 10 c and then subtracts from the dividend the product of 10 c by the divisor. The old dividend is now replaced by the remainder after this subtraction and al-Samaw'al says, "We copy the divisor one order to the right, as we do in the Indian style of arithmetic, and obtain the second chart".

The reader no doubt recognizes the procedure from the division of two whole numbers as Kūshyār ibn Labbān explains it, with its key sequence: divide leading term by leading term, multiply, subtract, and shift to the right, but this time the method is applied to divide algebraic expressions.

Now the sequence is repeated. In the Chart 4.2 we divide the leading 2 of the new dividend by the 2 of the divisor, and the quotient, 1, we place in the column to the right of the 10, in the row of the answer. Now we can forget which order the 2 of the divisor represents, for the chart keeps track of it for us. Then "1 times the divisor" is subtracted from the dividend, and the divisor is again shifted one place to the right to obtain the third chart.

Three more iterations of the procedure result in the last chart, where clearly the last row divides the row above it exactly, with a quotient of 2, and thus al-Samaw'al obtains the quotient of

$$20x^6 + 2x^5 + 58x^4 + 75x^3 + 125x^2 + 196x + 94 + 40x^{-1} + 50x^{-2} + 90x^{-3} + 20x^{-4}$$

by

$$2x^6 + 5x^4 + 5x^3 + 10x^{-1}$$

as

$$10x^3 + x^2 + 4x + 10 + 8x^{-2} + 2x^{-3}.$$

Our present method of division attaches labels (x, x^2, etc.) to the coefficients each time they are written down in order to keep track of what they represent; but, al-Samaw'al's method labels the columns of an array once and then arranges the coefficients in the labelled columns.

The Second Example

He follows this example of division by another example, in which he divides expressions involving negative coefficients and makes good use of his rules for signed quantities, including a subtraction $-20 - (-40) = 20$ of the sort that gave al-Karajī troubles.

The following, final, example illustrates al-Samaw'al's insight. The problem is that of dividing 20 m + 30 t by 6 m + 12 units, and though we shall again give all his charts combined into one we will not repeat the explanation of the procedure (Chart 4.3).

Al-Samaw'al calls the result so far, namely

$$3\tfrac{1}{3} + 5\cdot\frac{1}{x} - 6\tfrac{2}{3}\cdot\frac{1}{x^2} - 10\cdot\frac{1}{x^3} + 13\tfrac{1}{3}\cdot\frac{1}{x^4} + 20\cdot\frac{1}{x^5} - 26\tfrac{2}{3}\cdot\frac{1}{x^6} - 40\cdot\frac{1}{x^7},$$

"the answer approximately", and he then checks his work by writing out the product of this expression by the divisor and subtracting the product from the dividend to verify that the difference is the remainder $320x^{-6} + 480x^{-7}$.

By this point al-Samaw'al has seen the law that governs the formation of the coefficients of the quotient, for without further calculations he now writes down the coefficients of all succeeding powers of x up to that of x^{-28}, which he correctly gives as $54{,}613\frac{1}{3}$ (where the fractional parts of the numbers are written out in words). The law is, namely, that if a_{-n} is the nth coefficient then $a_{-n} = -\frac{1}{2}a_{-n-2}$.

māl	thing	unit	pt	pm	pc	pmm	pmc	pcc	pmmc	pmcc
		$3\frac{1}{3}$	5	$-6\frac{2}{3}$	-10	$13\frac{1}{3}$	20	$-26\frac{2}{3}$	-40	
20 6	30 0	 12								
	30 6	-40 0	 12							
		-40 6	-60 0	 12						
		-60 6	80 0	 12						
			80 6	120 0	 12					
				120 6	-160 0	 12				
					-160 6	-240 0	 12			
						-240 6	320 0	 12		
							320 6	480 0	 12	

Chart 4.3

However, apart from the insight such calculations show, the discovery of this procedure of long division, which is in all its computations precisely our present-day one, is a fine contribution to the history of mathematics, and it seems to be a joint accomplishment of al-Karajī and al-Samaw'al.

§7. ʿUmar al-Khayyāmī and the Cubic Equation

The Background to ʿUmar's Work

ʿUmar's treatment of cubic equations is found in his book *Algebra*, which he completed in Samarqand and dedicated to the chief judge of that city, Abū Ṭāhir. In the preface to this work he refers to his harried existence up to then:

> I have always desired to investigate all types of theorems . . . , giving proofs for my distinctions, because I know how urgently this is needed in the solution

of difficult problems. However, I have not been able to find time to complete this work, or to concentrate my thoughts on it, hindered as I have been by troublesome obstacles.

When ʿUmar did have the security to concentrate on a problem his powers of intellect were remarkable. One of his biographers, al-Bayhaqī, tells how ʿUmar was able to read a book seven times in Isfahan and memorized it. When he returned he wrote it out from memory, and subsequent comparison with the original revealed very few discrepancies. However, ʿUmar's feats of intellect were by no means confined to a remarkable memory, as we shall see in the sections from his great work, *Algebra*.

In his introduction to this work ʿUmar mentions that no algebraic treatment of the problems he is going to discuss has come down from the ancients, but that among the modern writers Abū ʿAbdallāh al-Māhānī wrote an algebraic analysis of a lemma Archimedes used in the problem from his work *Sphere and Cylinder* II,4 which we mentioned earlier, the problem of cutting a sphere by a plane so that the volumes of the two segments of the sphere are to one another in a given ratio. Archimedes showed that this problem can be solved if a line segment a can be divided into two parts b and c so that c is to a given length as a given area is to b^2. If we let $b = x$, so $c = a - x$, the proportion may be written as $x^3 + m = nx^2$, where m is the product of the given length and area. Khayyām tells us that neither al-Māhānī, who lived from 825–888 and would therefore have been contemporary with al-Khwārizmī, nor Thābit, could solve this equation, but a mathematician of the next generation, Abū Jaʿfar al-Khāzin, did solve it by means of intersecting conic sections. Then, following Abū Jaʿfar, various mathematicians tried to solve special kinds of these equations involving cubes, but no one had tried to enumerate all possible equations of this type and solve them all. This ʿUmar says, he will do in this treatise.

ʿUmar's Classification of Cubic Equations

We now give an account of some parts of his treatise *Algebra*, and we emphasize that although we shall speak of "equations" and "coefficients" ʿUmar did not write these symbolically, for he used only words, even for the numbers.

In the first part of his treatise ʿUmar lists all types of equations in which no term of degree higher than three occurs. In ʿUmar's equations all terms appear with positive coefficients so that, whereas we would see $x^3 - 3x + 8 = 0$ and $x^3 + 3x - 8 = 0$ as being of the same type, ʿUmar viewed them as being different. He would have expressed the first as "Cube and numbers equal sides" ($x^3 + 8 = 3x$) and would have seen it as distinct from "Cube and sides equal numbers" ($x^3 + 3x = 8$). Thus, he arrives at 25 species of equations, and, in the remainder of the treatise, he shows how these may be solved—11 by Euclidean methods and 14 by conic sections. For each of these 14 species

ʿUmar gives a short section showing how the conics can be used to produce a line segment from which solids that satisfy the required relation can be constructed. The student who reads the English translation by D. S. Kasir will find ʿUmar's arguments quite clear and will enjoy, as well, the many interesting asides on the history of various types of equations.

ʿUmar's Treatment of $x^3 + mx = n$

Preliminaries

The equation we are going to discuss is "cube and sides equal a number", that is, "$x^3 + mx = n$", and, to understand the section that we have chosen to present from this work, the reader must recall that if ABC is a parabola with vertex B and parameter p and if x is any abscissa and y the corresponding ordinate then $y^2 = p \cdot x$.

ʿUmar begins with a lemma about solid figures called parallelopipeds, solids with three pairs of parallel faces (Fig. 4.8(a)). When all the faces are rectangles (Fig. 4.8(b)), as in the case of a brick, the solid is called a rectangular parallelopiped. One face of a parallelopiped is arbitrarily designated as its base, and ʿUmar's lemma concerns the case in which the base is a square.

Lemma. *Given a rectangular parallelopiped* ABGDE (Fig. 4.9), *whose base is the square* ABGD $= a^2$ *and whose height is* c, *and given another square* MH $= b^2$, *construct on* MH *a rectangular parallelopiped equal to the given solid.*

Solution. Use Euclidean geometry to construct a line segment k so that $a:b = b:k$; and then construct h so that $a:k = h:c$. Then the solid whose base is b^2 and whose height is h is equal to the given solid.

Proof. $a:b = b:k$ implies $a^2:b^2 = (a:b)\cdot(a:b) = (a:b)\cdot(b:k) = a:k$, but $a:k = h:c$, and so $a^2:b^2 = h:c$, and this means $a^2 \cdot c = b^2 \cdot h$. Thus the solid

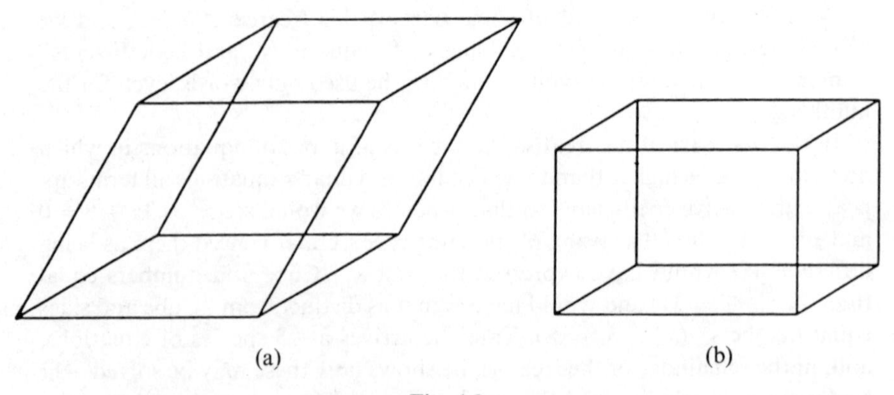

Fig. 4.8

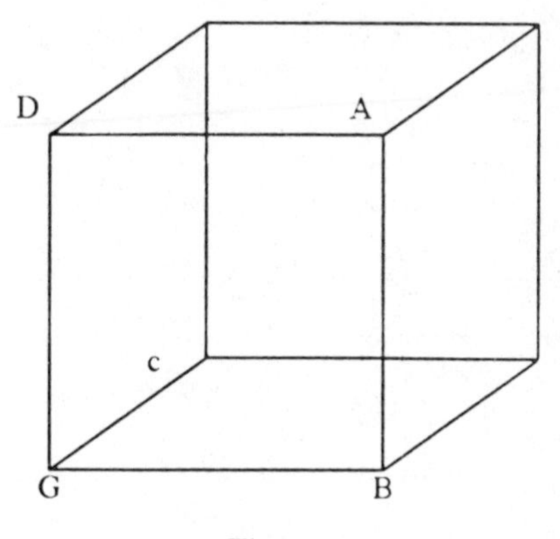

Fig. 4.9

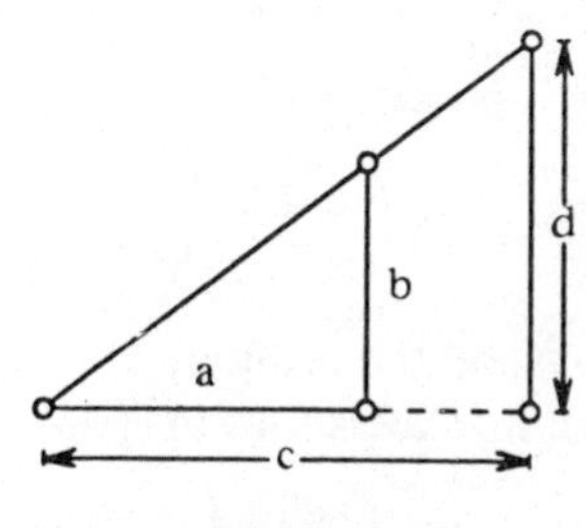

Fig. 4.10

whose base is b^2 and whose height is h is equal to the given solid $a^2 \cdot c$, and that is what we wanted to show.

Algebraically this lemma asks for the root of $a^2 \cdot c = b^2 \cdot x$, given a, c and b, and Khayyām's constructions obtain this solution $(a^2 \cdot c/b^2)$ by first obtaining $k = b^2/a$ and then $h = (ac)/k = a^2 \cdot c/b^2$. The principal fact, taken as known by ʿUmar, is that given any three straight-line segments a, b, c it is possible to find a fourth segment d so that $a:b = c:d$. The segment d is called "the fourth proportional", and Fig. 4.10 shows how d can be constructed.

The Main Discussion

ʿUmar now comes to his first nontrivial equation, which he describes as "Cube and sides equal a number", i.e. the case we would write as $x^3 + mx = n$, where m and n are positive. For this he gives the following procedure: Let b be the side of a square that is equal to the number of roots, i.e. $b^2 = m$, and let h be the height of the rectangular parallelopiped whose base is b^2 and whose volume is n. (The construction of h follows immediately from the previous lemma.) Now take a parabola (Fig. 4.11) whose vertex is B, axis BZ and parameter b, and place h perpendicular to BZ at B. On h as diameter describe a semicircle and let it cut the parabola at D. From D

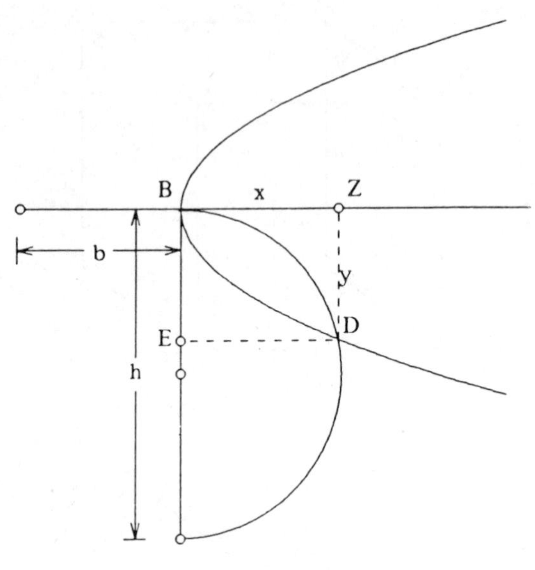

Fig. 4.11

drop DE perpendicular to h and the ordinate DZ perpendicular to BZ. Then DZ = EB and with y = BE it follows that $y^3 + my = n$.

Proof. Let BZ $= x$. By the properties of the parabola $y^2 = bx$ and, by the properties of the circle, $x^2 = y(h - y)$. But the first equality may be written as $x:y = y:b$ and the second as $x:y = (h - y):x$. Thus $(h - y):x = x:y = y:b$, or, inverting, $b:y = y:x = x:(h - y)$. Hence $b^2:y^2 = (b:y)(b:y) = (y:x)(x:(h - y)) = y:(h - y)$, and thus $b^2 \cdot (h - y) = y \cdot y^2$, i.e. $b^2 \cdot h - b^2 \cdot y = y^3$. Therefore, if we add $b^2 \cdot y$ to both sides it follows that $b^2 \cdot h = y^3 + b^2 \cdot y$. If we then substitute for $b^2 \cdot h$ its equal, n, and for b^2 its equal, m, we may conclude that $y^3 + my = n$, which was the equation we wanted to solve.

We can write the argument a bit more briefly as $y^2 = bx$ implies $y^4 = b^2x^2$, and $x^2 = y(h - y)$ implies $b^2x^2 = b^2y(h - y)$. Thus, $y^4 = b^2x^2 = b^2y(h - y)$ and so, because $y \neq 0$,

$$y^3 = b^2(h - y), \quad \text{i.e.} \quad y^3 + my = y^3 + b^2 \cdot y = b^2 \cdot h = n$$

and the equation is solved.

ʿUmar's Discussion of the Number of Roots

Throughout the discussion ʿUmar is careful to warn the reader that a particular case may have more than one solution (or, as we should say, more than one positive real root) or that it may have no solutions. What happens in any given case depends on whether the conic sections he is using intersect in none, in one or in two points. For example, he obtains the solution to $x^3 + n = mx$ by intersecting a parabola and hyperbola and notices that the two curves may not intersect, in which case there would be no solution, but if they do then they either intersect tangentially or at two points. In our modern terminology we would express this by saying the equation $x^3 + n = mx$ either has no positive real solution or two of them. In the latter case the two could be a single root, a repeated root, corresponding to a factor $(x - a)^2$, or two different roots. Again, in the case of $x^3 + n = mx^2$ he notes that if $\sqrt[3]{n} \geq m$ then there is no solution. For, if $\sqrt[3]{n} \geq m$ then

$$n = (\sqrt[3]{n})^3 = \sqrt[3]{n}(\sqrt[3]{n})^2 \geq m(\sqrt[3]{n})^2. \tag{1}$$

This implies that if x is any solution then $x > \sqrt[3]{n}$ for

$$x^3 + n = mx^2 \quad \text{implies} \quad mx^2 > n. \tag{2}$$

Thus, combining (1) and (2) we obtain

$$mx^2 > n \geq m(\sqrt[3]{n})^2, \quad \text{so} \quad mx^2 > m(\sqrt[3]{n})^2, \quad \text{i.e.} \quad x > \sqrt[3]{n}.$$

Thus $\sqrt[3]{n} \geq m$ implies for any solution x, $x > \sqrt[3]{n}$.

On the other hand, for any solution, $x^3 < mx^2$, so that $x < m < \sqrt[3]{n}$, and this contradicts $x > \sqrt[3]{n}$. Hence ʿUmar has shown if $\sqrt[3]{n} \geq m$ then $x^3 + n = mx$ has no (positive real) solution.

The reader should be aware that we have presented ʿUmar's argument by means of our modern symbolic algebra, a product of the European Renaissance. ʿUmar's argument uses either geometrical magnitudes or numbers interpreted geometrically, and the only mathematical symbols he uses are letters to denote points on geometrical diagrams. Thus, for example, in the case of the preceding argument, ʿUmar expresses the condition $\sqrt[3]{n} = m$ by saying, "Let AC represent the number (m) of squares and describe a cube equal to the given number (n), the side of which is h .... If h is equal to AC the problem will be impossible because ...". (ʿUmar's pleasure at having discovered that certain cases that earlier workers had thought impossible are in fact quite possible, is evident at several times in the treatise.)

However, having stressed the geometrical form of ʿUmar's argument, we must emphasize also that ʿUmar looked on his work as being a contribution to algebra. After the customary invocation of Allah and prayers for His blessings on His prophet Muḥammad he says, "One of the branches of knowledge needed in that division of philosophy known as mathematics is the science of algebra, which aims at the determination of numerical and geometrical unknowns ...." Further he begins his first chapter by saying,

> Algebra. By the help of God and with His precious assistance I say that algebra is a scientific art. The objects with which it deals are absolute numbers and (geometrical) magnitudes which, though themselves unknown, are related

to things which are known, whereby the determination of the unknown quantities is possible.... What one searches for in the algebraic art are the relations that lead from the known to the unknown, to discover which is the object of algebra as stated above.

If these relations that lead us from known to unknown happen to stem from the properties of geometric figures the problem is no less algebraic in ʿUmar's view. It is the use of the given relations to search for the unknown that is the hallmark of algebra—nothing else.

In fact, even though ʿUmar's treatise expresses the solutions as line segments and not as numbers depending on the coefficients of the equation, we know ʿUmar wanted to find such numbers for he writes: "As for a demonstration of these types, if the object of the problem is an absolute number, neither we nor any of the algebraists have succeeded, except in the case of the first three degrees, namely number, thing and square, but maybe those after us will."

ʿUmar's hope was prophetic, for in the early part of the sixteenth century in Italy a group of algebraists put together different pieces of the puzzle, and in 1545 the physician and astrologer Gerolamo Cardano published in his *Ars Magna* (*The Great Art*), just as ʿUmar had, a case-by-case analysis of the cubic equation, but this time the roots were expressed not as line segments but as numbers depending on the coefficients of the equation. Instead of conic sections, Cardano employed identities like $(a - b)^3 + 3ab(a - b) = a^3 - b^3$ to show how the solution of the cubic could be obtained from a solution of an associated quadratic followed by the extraction of cube roots—both of which could have been done numerically long before ʿUmar's time. The formula Cardano published for a root of the equation $x^3 + px = q$ is:

$$x = \sqrt[3]{\frac{q}{2} + \sqrt{\left(\frac{q}{2}\right)^2 + \left(\frac{p}{3}\right)^3}} - \sqrt[3]{-\frac{q}{2} + \sqrt{\left(\frac{q}{2}\right)^2 + \left(\frac{p}{3}\right)^3}}$$

and in this way ʿUmar's hopes were fulfilled, over four centuries later.

§8. The Islamic Dimension: The Algebra of Legacies

To illustrate further mathematical methods in the service of Islam we turn, as in the chapter on arithmetic, to the latter half of al-Khwārizmī's *Algebra* and a problem from the science of legacies (*ʿilm al-waṣāyā*). This science, the reader may recall, requires the application of religious law and algebra to calculating shares of an estate when a legacy to a stranger is involved.

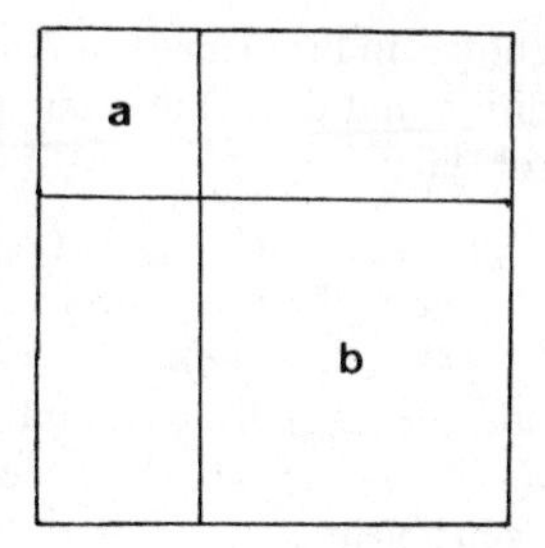

Fig. 4.12

Al-Khwārizmī gives the following example of the use of algebra in the case that there is a legacy:

> A man dies, leaving two sons and bequeathing one-third of his estate to a stranger. His estate consists of ten *dirhams* of ready cash and ten *dirhams* as a claim against one of the sons, to whom he has loaned the money.

The relevant parts of Islamic inheritance law are:

(1) The natural heirs can only refuse to pay that part of a legacy by which it exceeds one-third of the estate, so in this case the legacy must be paid;
(2) The amount by which the outstanding loan to the son exceeds what would be his legal share is treated as a gift to the son; and
(3) The gift precedes the legacy and the legacy precedes the natural shares.

To solve this problem, al-Khwārizmī lets x be the legal share of each son. Since the gift must be paid first, only the legal share is taken out of the 10 *dirham* loan and added to the 10 *dirhams* of ready cash to form the estate $10 + x$. The stranger than gets one-third of this, and each son gets x, so $(10 + x)/3 + 2x = 10 + x$. Thus $x = (10 + x)/3$, so that $\frac{2}{3}x = 3\frac{1}{3}$. Then, when each side is increased by half, $\frac{2}{3}x$ becomes x and $3\frac{1}{3}$ becomes 5. (Recall that arithmetics had separate sections on halving a number.) Thus the legal share is 5, the stranger gets $(10 + 5)/3 = 5$, and the gift is 5.

Exercises

1. Show that the fourth type of quadratic equation, $x^2 + n = mx$, is the only one of al-Khwārizmī's six types that can have two positive solutions.
2. Find a proof similar to the first case in the case of the second figure of *Euclid* II,5.
3. With reference to Thābit's discussion of the second basic form of the quadratic equation, show that if $a^2 + q = pa$, then p is greater than a.
4. Make use of Fig. 4.12 to show that $\sqrt{a} + \sqrt{b} = \sqrt{a + b + 2\sqrt{ab}}$ and draw your own diagram to prove that $\sqrt{a} - \sqrt{b} = \sqrt{a + b - 2\sqrt{ab}}$.

5. Verify that the three quantities in Problem 61 of Abū Kāmil's book, b, yb and zb, not only satisfy the condition that their sum is ten, but they also satisfy the other two conditions of the problem.
6. Suppose $f(x, y, \ldots, w)$, $g(x, y, \ldots, w)$, $\ldots$, $k(x, y, \ldots, w)$ are homogeneous polynomials and f is of degree 1. If c is a nonzero constant show that the conditions $f(1, Y, \ldots, W) = c$, $g(1, Y, \ldots, W) = \cdots = k(1, Y, \ldots, W) = 0$ are satisfied if and only if, with $x = (d/c)X$, $y = (d/c)Y$, etc., $f(x, y, \ldots, w) = d$ and $g(x, y, \ldots, w) = \cdots = k(x, y, \ldots, w) = 0$. Also identify all these functions and constants in Abū Kāmil's problem.
7. Show that the root Abū Kāmil found for the equation arising from Problem 61 is positive, but less than 5, while the other root of the equation exceeds 5. Deduce that the root Abū Kāmil has located is the correct value of $b = 10/a$.
8. Use al-Samaw'al's procedure for division to find the quotient of $2x^3 - 11x^2 - 13x - 5$ by $2x - 5$.
9. Solve the following problem from al-Khwārizmī. A man has two sons and leaves ten *dirhams* in ready cash and ten *dirhams* as a claim against one of the sons, to whom he has loaned the money. To a stranger he leaves one-fifth of his estate plus one *dirham*. Compute the amount each party will receive.

Bibliography

Abū Kāmil and Shujā'b. Aslam, *Algebra* (transl. and comm. by M. Levey). Madison, WI: University of Wisconsin Press, 1966.

Gandz, S. "The Algebra of Inheritance", *Osiris* **5** (1938), 319–391.

al-Khayyāmī, 'Umar. *The Algebra* (transl. and comm. by D. S. Kasir as *The Algebra of Omar Khayyam*). New York: 1931.

Another English translation is by H. J. J. Winter and W. 'Arafat. "The Algebra of 'Umar Khayyam", *Journal of the Royal Asiatic Society of Bengal. Science* **16** (1950), 27–70.

al-Khwārizmī, Muḥammad b. Mūsā, (transl. by F. Rosen). *The Algebra of Muhammed ben Musa*. London: 1831.

Rashed, R. "Récommencements de l'Algèbre au XIe et XIIe Siècles". In: *The Cultural Context of Medieval Learning* (J. E. Murdoch and E. D. Sylla, eds.). Dordrecht: Reidel, 1975, pp. 33–60.

Sesiano, J. "Les Méthodes d'analyse indéterminée chez Abū Kāmil", *Centaurus* **21** (No. 2) (1977), 89–105.

Chapter 5
Trigonometry in the Islamic World

§1. Ancient Background: The Table of Chords and the Sine

The branch of elementary mathematics whose origins most clearly lie in astronomy is trigonometry, for there is no trace of this subject until Hellenistic astronomers devised models for the motion of the sun, moon and five known planets which required calculating the values of certain sides and angles of a triangle from other, given, ones. This happened at least as early as the time of Apollonios of Perga in the third century B.C. Astronomers of ancient India also used the Greek models and therefore faced the same mathematical problems, and it is the astronomical handbooks, or commentaries on them, by Greek and Indian authors that furnish most of our record of the early history of trigonometry.

Of course, it is a poor problem that inspires only one solution, and the problems of astronomy, being very good problems indeed, drew forth a wonderful variety of solutions which ranged from the construction of numerical sequences to the methods of descriptive geometry. Among these methods were those that we recognize as being trigonometric, and in order to provide a historical context for the Islamic contributions we turn to a brief survey of the developments in Greece and India.

The most thorough Greek discussion of trigonometry is contained in *The Almagest*, which is, as we have mentioned earlier, an astronomical work written in the early part of the second century A.D. by the astronomer Ptolemy who worked in Alexandria. The word "Almagest", is an Arabic rendering of a Greek word *megistē* (with the Arabic definite article *al-* put in front) meaning "the greatest", for Greek writers called Ptolemy's *Mathematical Arrangement* by the name *The Great Arrangement*, and so important did his work become that later astronomers, perhaps punning on the dual sense of a Greek word that could mean either "big" or "great", called it simply "The Greatest". Islamic writers transliterated the Greek into Arabic and in the twelfth-century European authors Latinized the Arabic to give us "Almagest".

In *The Almagest* Ptolemy supplies tables and rules to allow the user to answer, for each of the sun, moon or five planets then known, the question, "Where will it be at a given time?". So that the user may understand how these tables and rules were composed, Ptolemy constructs the underlying geometrical models, whose parameters he derived from observations by ingenious mathematical methods. These models were so successful that they formed the basis for scientific astronomy to the time of Copernicus.

Many of Ptolemy's mathematical methods depend on a table which he places in Book I of *The Almagest* and which he calls "A TABLE OF CHORDS IN A CIRCLE". In Fig. 5.1 we have reproduced the three columns of part of the table. The left-most column simply lists the arcs of a circle, beginning with an arc of $\frac{1}{2}^\circ$ and proceeding by steps of $\frac{1}{2}^\circ$ up to 180°. The next column lists, for each arc θ, the length of the chord that subtends the arc in a circle of radius 60. The length is given, using sexagesimal fractions, in terms of the 60 units in the radius. The third column tells the average increase in the chord length per minute of arc, from one entry to the next, and is used for purposes of linear interpolation. Thus, for example, row 3 reads "Crd $1\frac{1}{2}^\circ = 1;34,15$ and for each of the next $30'$ of arc add 0;1,2,50 to the chord length."

A table such as this, together with a knowledge of how to use it, was all the trigonometry that the ancient Hellenistic astronomer had available to solve plane trigonometric problems. However, this one table was powerful enough so that even as late as the thirteenth century the Muslim astronomer Naṣīr al-Dīn al-Ṭūsī discusses its use in his great work called *The Transversal Figure*.

In Chapter 2 of Part III Naṣīr al-Dīn begins by remarking that since any triangle ABG may be inscribed in a circle its sides may be regarded as the chords subtending its angles, or, more properly, subtending the arcs opposite these angles (Fig. 5.2). He then says that both in astronomical operations and in geometry itself it is necessary to find some angles and sides of triangles from other given ones, and this may be done either by arcs and chords or arcs and sines. Our immediate concern is with his use of arcs and chords to solve right triangles. (We have *solved* a triangle when we have obtained the values of all its sides and angles.)

Naṣīr al-Dīn remarks, first of all, that if only one acute angle of a right triangle ABG is known then all the angles are known, for the two acute angles must together total 90°; however, only the ratios of the sides will be known. Here he is doubtless referrring to Prop. VI,4 of Euclid's *Elements*, which says that if two triangles have equal angles then the sides about equal angles are proportional.

Thus one must know at least one side, and Naṣīr al-Dīn first assumes he knows two (say a, g or a, b in Fig. 5.3), so he can calculate the third from the Pythagorean theorem. To calculate the magnitude of an acute angle in this

TABLE OF CHORDS

Arcs	Chords	Sixtieths	Arcs	Chords	Sixtieths
½	0 31 25	1 2 50	23	23 55 27	1 1 33
1	1 2 50	1 2 50	23½	24 26 13	1 1 30
1½	1 34 15	1 2 50	24	24 56 68	1 1 26
2	2 5 40	1 2 50	24½	25 27 41	1 1 22
2½	2 37 4	1 2 48	25	25 58 22	1 1 19
3	3 8 28	1 2 48	25½	26 29 1	1 1 15
3½	3 39 52	1 2 48	26	26 59 38	1 1 11
4	4 11 16	1 2 47	26½	27 30 14	1 1 8
4½	4 42 40	1 2 47	27	28 0 48	1 1 4
5	5 14 4	1 2 46	27½	28 31 20	1 1 0
5½	5 45 27	1 2 45	28	29 1 50	1 0 56
6	6 16 49	1 2 44	28½	29 32 18	1 0 52
6½	6 48 11	1 2 43	29	30 2 44	1 0 48
7	7 19 33	1 2 42	29½	30 33 8	1 0 44
7½	7 50 54	1 2 41	30	31 3 30	1 0 40
8	8 22 15	1 2 40	30½	31 33 50	1 0 35
8½	8 53 35	1 2 39	31	32 4 8	1 0 31
9	9 24 54	1 2 38	31½	32 34 22	1 0 27
9½	9 56 13	1 2 37	32	33 4 35	1 0 22
10	10 27 32	1 2 35	32½	33 34 46	1 0 17
10½	10 58 49	1 2 33	33	34 4 55	1 0 12
11	11 30 5	1 2 32	33½	34 35 1	1 0 8
11½	12 1 21	1 2 30	34	35 5 5	1 0 3
12	12 32 36	1 2 28	34½	35 35 6	0 59 57
12½	13 3 50	1 2 27	35	36 5 5	0 59 52
13	13 35 4	1 2 25	35½	36 35 1	0 59 48
13½	14 6 16	1 2 23	36	37 4 55	0 59 43
14	14 37 27	1 2 21	36½	37 34 47	0 59 38
14½	15 8 38	1 2 19	37	38 4 36	0 59 32
15	15 39 47	1 2 17	37½	38 34 22	0 59 27
15½	16 10 56	1 2 15	38	39 4 5	0 59 22
16	16 42 3	1 2 13	38½	39 33 46	0 59 16
16½	17 13 9	1 2 10	39	40 3 25	0 59 11
17	17 44 14	1 2 7	39½	40 33 0	0 59 5
17½	18 15 17	1 2 5	40	41 2 33	0 59 0
18	18 46 19	1 2 2	40½	41 32 3	0 58 54
18½	19 17 21	1 2 0	41	42 1 30	0 58 48
19	19 48 21	1 1 57	41½	42 30 54	0 58 42
19½	20 19 19	1 1 54	42	43 0 15	0 58 36
20	20 50 16	1 1 51	42½	43 29 33	0 58 31
20½	21 21 11	1 1 48	43	43 58 49	0 58 25
21	21 52 6	1 1 45	43½	44 28 1	0 58 18
21½	22 22 58	1 1 42	44	44 57 10	0 58 12
22	22 53 49	1 1 39	44½	45 26 16	0 58 6
22½	23 24 39	1 1 36	45	45 55 19	0 58 0

Fig. 5.1

Taken from Toomer: Ptolemy's *Almagest*, copyright by Springer-Verlag 1984, reproduced by permission of Springer-Verlag.

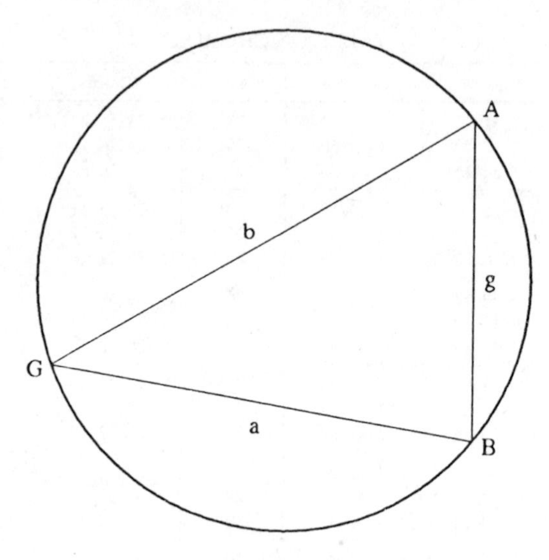

Fig. 5.2

case imagine a new unit, u, so that $b = 120u$, and re-calculate the sides to this scale. (This change of scale to a diameter of 120 units runs all through ancient trigonometry, because the tables of chords and sines were constructed on the assumption of this diameter.) If $a = n \cdot u$ then one may look down the column headed "Chords" in a table of chords in a circle to find the entry n, and the entry next to it will be the magnitude of $\widehat{BG}$. From this, $\sphericalangle A = \frac{1}{2}\widehat{BG}$, and as before, $\sphericalangle G$ can be calculated since $\sphericalangle A + \sphericalangle G = 90°$.

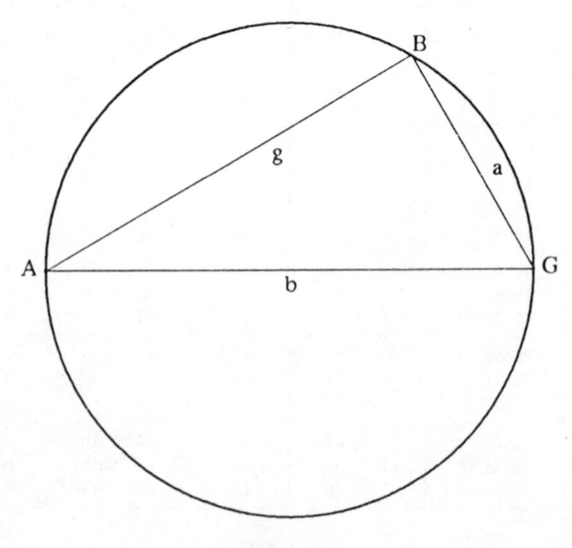

Fig. 5.3

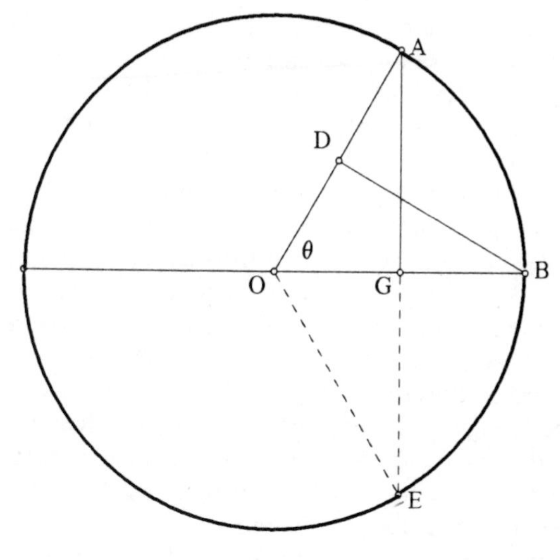

Fig. 5.4

In the third case, when one is given only one side he must also know an acute angle A. Then $\measuredangle B$ may be calculated as $90° - \measuredangle A$, so that all angles are known. Again, it is only changing scale to assume the hypotenuse has 120 units and set up the proportion involving the known side and either unknown side:

$$GB:GA = \text{Crd}(\widehat{GB}):\text{Crd}(\widehat{GA}).$$

All arcs are known since each is twice the opposite known angle, and one of the two sides is known. Thus we may calculate the other side from the above relation, and this concludes Naṣīr al-Dīn's presentation of how to solve any right triangle by means of a table of chords.

Thus a table of chords in a circle may be used in place of a sine table to solve triangles. Indeed, since in Fig. 5.4,

$$\sin\theta = \frac{AG}{AO} = \frac{\frac{1}{2}\,\text{Crd}(2\widehat{AB})}{60} = \frac{\frac{1}{2}\,\text{Crd}(\widehat{AE})}{60},$$

a table of chords for $\widehat{AE}$ from 0° to 180° is equivalent to a table of sines for θ from 0° to 90°. In addition, relations like $\cos(\theta) = \sin(90° - \theta)$ and $\tan\theta = \sin\theta/\cos\theta$ show how all trigonometric functions can be expressed in terms of the sine function, so a single table of chords can be made to serve for all the trigonometric functions. Thus, with such a table we can determine three unknown quantities in a triangle from three known quantities, whenever this is possible with trigonometric functions.

Of course, what is possible in theory may be awkward in practice, for one function is usually no substitute for six, each one of which is tailor-made for a certain kind of application. In addition, as we saw in the example, there is often the inconvenience of having to double an angle in order to calculate the size of an arc from that of an inscribed angle. However, so far as we know, these inconveniences never motivated the Hellenistic astronomers to abandon a tool that was used as early as the second century B.C. by the astronomer Hipparchos on the island of Rhodes.

It is, rather, the astronomers of India who introduced the function that is basic to modern trigonometry, namely the sine, and in an Indian astronomical handbook dating from the fourth or fifth centuries of our era and known as the *Surya Siddhanta*, the values of this function are given in Sanskrit verse for every $3\frac{3}{4}°$ of arc, from $3\frac{3}{4}°$ to $90°$. The radius of the reference circle was taken to be $3438'$, where the minute is a unit of length equal to $\frac{1}{60}$ of the length of $1°$ of arc on the circle. With this Hindu achievement we have come near to the beginnings of the Islamic reception of Indian astronomy, for it was in the eighth century, at about the time Islam had spread from Spain to China, that Indian astronomical texts were translated into Arabic at the court of the caliph al-Manṣūr in Baghdad.

Soon Muslim astronomers were producing their own astronomical handbooks, based on Indian and Greek models but incorporating elements of particular concern to Islamic civilization, e.g. visibility of the crescent moon and the direction of Mecca. Such handbooks, written by almost all important Islamic astronomers, were called *zījes*, a Persian word taken over into Arabic. The word originally signified a "thread" or "chord", and then a set of these, as in the warp of a fabric. By analogy, astronomical tables, presenting the appearance of a whole set of parallel lines separating the columns, came to be known by the same word.

§2. The Introduction of the Six Trigonometric Functions

Among the ways in which authors of the Islamic world extended the ancient methods in trigonometry was to define and use all six trigonometric functions, as follows:

(1) *The Sine* (Fig. 5.4). This was defined for an arc $\widehat{AB}$ of a circle with center O and radius R as the length of the perpendicular AG from A onto OB. Clearly, this is also the length of the perpendicular BD from B onto OA, and its length depends on R. In fact, if $\text{Sin}_R\,\widehat{AB}$ denotes the sine of an arc AB in a circle of radius R, then this medieval sine function is related to the modern function by the rule $\text{Sin}_R\,\text{AB} = R \cdot \sin \widehat{AB}$, and to Ptolemy's chord function as $\text{Sin}_R\,\widehat{AB} = \frac{1}{2}\,\text{Crd}_R\,2\widehat{AB}$. (We shall use capital letters to denote the Indian and small letters the modern trigonometric functions.)

(2) *The Cosine.* Again, Muslim authors described this as a length and not a ratio. If $\text{Cos}_R\,\widehat{AB}$ denotes this function for arcs $\widehat{AB} < 90°$ in a circle of

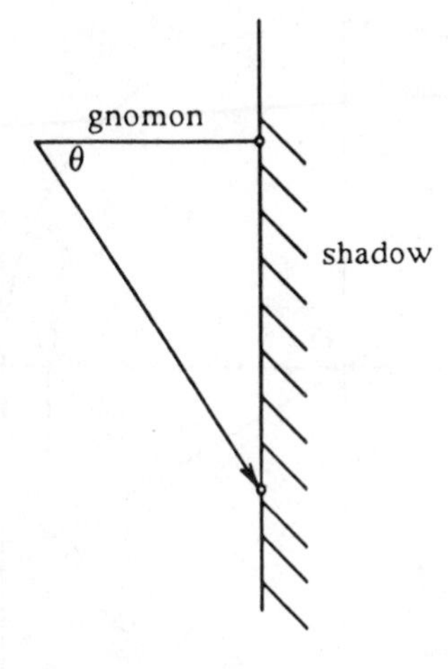

Fig. 5.5

radius R then

$$\mathrm{Cos}_R\, \widehat{\mathrm{AB}} = \mathrm{Sin}_R(90^\circ - \widehat{\mathrm{AB}}),$$

and this is the length OG in Fig. 5.4. The function was always called "the Sine of the complement of the arc" and was not tabulated separately.

(3) and (4) *The Tangent and Cotangent*. Both were originally conceived of as lengths of certain shadows, the tangent being the shadow of a horizontal rod mounted on a wall (for a given altitude of the sun) and the cotangent the shadow of a vertical rod (*gnōmōn* in Greek and *miqyās* in Arabic) of standard length. Thus, in Figs. 5.5 and 5.6, the shadow lengths are, respectively, $R \cdot \tan \theta$ and $R \cdot \cot \theta$, where R is the length of the rod and θ the angle of elevation of the sun over the horizon.

However, by the tenth century both functions were defined, as we find them in Naṣīr al-Dīn's work, according to Fig. 5.7. Here BD is perpendicular to OB, AG is perpendicular to OB, and EK is perpendicular to EO. Then

$$\mathrm{Tan}_R\, \widehat{\mathrm{AB}} = \mathrm{DB} \quad \text{and} \quad \mathrm{Cot}_R\, \widehat{\mathrm{AB}} = \mathrm{EK}.$$

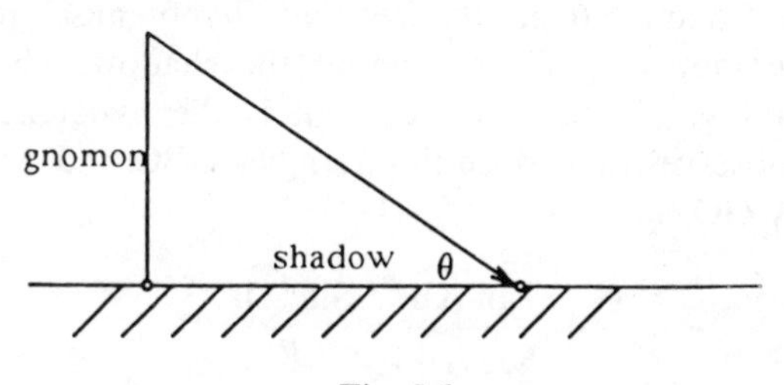

Fig. 5.6

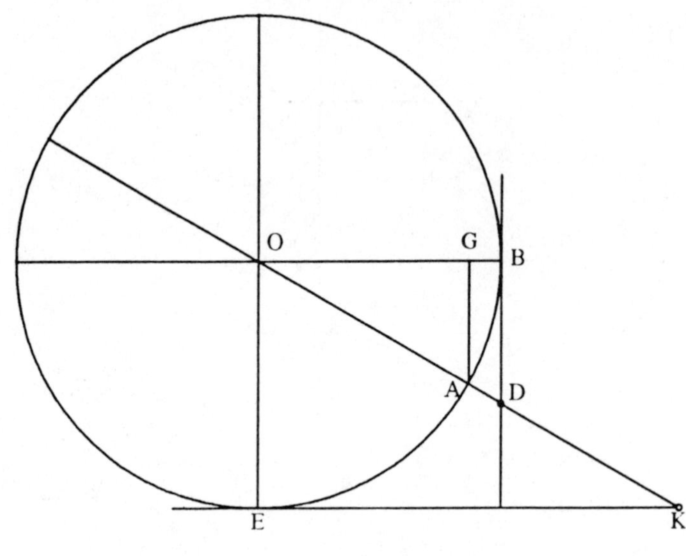

Fig. 5.7

Muslim authors tabulated the Tangent function, but since they recognized that the Cotangent is just the Tangent of the complement they did not tabulate it separately. Beyond the preceding relation, Naṣīr al-Dīn states the following pair of relations:

1. $\text{Tan}\,\widehat{AB}/R = \text{Sin}\,\widehat{AB}/\text{Cos}\,\widehat{AB}$. (Notice that when $R = 1$ all the functions become their modern counterparts and the relation is the familiar $\tan \widehat{AB} = \sin \widehat{AB}/\cos \widehat{AB}$.)
2. $\text{Tan}\,\widehat{AB}/R = R/\text{Cot}\,\widehat{AB}$. (When $R = 1$ this becomes $\tan \widehat{AB} = 1/\cot \widehat{AB}$, a relation familiar to us.)

(5) and (6) *The Secant and Cosecant*. These functions were seldom tabulated, but Naṣīr al-Dīn defines them, with reference to Fig. 5.7, as $\text{Sec}\,\widehat{AB} = \text{OD}$ and $\text{Csc}\,\widehat{AB} = \text{KO}$. In the terminology of the Arabic writers these were called "the hypotenuse of the shadow" and "the hypotenuse of the reversed shadow" respectively, and these names may be explained by reference to Figs. 5.5 and 5.6 and the fact that "hypotenuse" refers to the line joining the gnomon's tip to the end of the shadow. The Secant is the hypotenuse in Fig. 5.5 and the Cosecant is the hypotenuse in Fig. 5.6. Naṣīr al-Dīn observes that, since the triangles DBO and AGO are similar, DB/DO = GA/GO, so

$$\frac{\text{Tan}\,\widehat{AB}}{\text{Sec}\,\widehat{AB}} = \frac{\text{Sin}\,\widehat{AB}}{R}.$$

The completion of the ancient systems of trigonometry to one based on the six functions we now use made trigonometry much simpler, and therefore more useful, than it had been before Islamic times.

§3. Abu l-Wafā''s Proof of the Addition Theorem for Sines

We see from Naṣīr al-Dīn's discussion that Islamic astronomers were aware of six trigonometric functions, which are, in every case, constant multiples of the modern ones. Also, from the late tenth century onward, following the work of Abu l-Wafā', they were aware of the possibility of taking $R = 1$ in defining the Sine, Cosine and Tangent. Thus Abu l-Wafā' may be regarded as the first to have calculated the modern trigonometric functions, and the simplifications such a step yields may be seen by comparing two statements, each equivalent to the familiar addition theorem for sines:

$$\sin(a \pm b) = \sin a \cdot \cos b \pm \cos a \cdot \sin b.$$

The first is an ancient form of this law from Ptolemy's *Almagest*, where Ptolemy gives rules (see Fig. 5.8) for finding (1) $\text{Crd}(\widehat{AB} - \widehat{AC})$, from $\text{Crd}(\widehat{AB})$ and $\text{Crd}(\widehat{AC})$, and (2) $\text{Crd}(\widehat{AC} + \widehat{CB})$ from $\text{Crd}(\widehat{AC})$ and $\text{Crd}(\widehat{CB})$. These may be represented as

(1) $\text{Crd}(\widehat{AB} - \widehat{AC}) \cdot \text{Crd}(180°) = \text{Crd}(\widehat{AB}) \cdot \text{Crd}(180° - \widehat{AC}) - \text{Crd}(\widehat{AC}) \cdot \text{Crd}(180° - \widehat{AB})$; and

(2) $\text{Crd}(180° - \widehat{AB}) \cdot \text{Crd}(180°) = \text{Crd}(180° - \widehat{AC}) \cdot \text{Crd}(180° - \widehat{CB}) - \text{Crd}(\widehat{AC}) \cdot \text{Crd}(\widehat{AB})$.

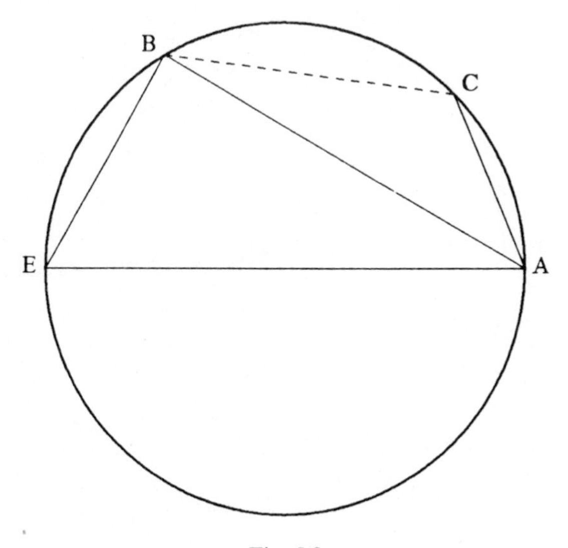

Fig. 5.8

In case (1), Crd(180°) = 120, $\text{Crd}(\overset{\frown}{AB})$, and $\text{Crd}(\overset{\frown}{AC})$ are all known and (see Fig. 5.8) $\text{Crd}(180° - \overset{\frown}{AC})$ and $\text{Crd}(180° - \overset{\frown}{AB})$ may both be calculated from these and the Pythagorean theorem since:

$$\text{Crd}(180° - \overset{\frown}{AC}) = \sqrt{(\text{Crd}(180°))^2 - (\text{Crd}(\overset{\frown}{AC}))^2},$$

and

$$\text{Crd}(180° - \overset{\frown}{AB}) = \sqrt{(\text{Crd}(180°))^2 - (\text{Crd}(\overset{\frown}{AC}))^2}.$$

Hence five of the six terms in (1) are known so we may solve for $\text{Crd}(\overset{\frown}{AB} - \overset{\frown}{AC})$. In Case (2) one solves for $\text{Crd}(180° - \overset{\frown}{AB})$ and from it calculates $\text{Crd}(\overset{\frown}{AB})$ by the Pythagorean theorem. Thus in both cases the desired quantities may be calculated, but with some effort.

Compare statements (1) and (2), which are not easy to remember, with the following elegant statement and unified proofs found in Abu l-Wafā''s *zīj Almajisṭī*.

> Calculation of the sine of the sum of two arcs and the sine of their difference when each of them is known. Multiply the sine of each of them by the cosine of the other, expressed in sixtieths, and we add the two products if we want the sine of the sum of the two arcs, but take the difference if we want the sine of their difference.

For this statement, which may be expressed by the modern formula given earlier, Abu l-Wafā' gives the following proof, which refers to Fig. 5.9 and 5.10. (Since he takes the radius of the circle to be unity we may write his trigonometric functions without capital letters. The reference to the trigo-

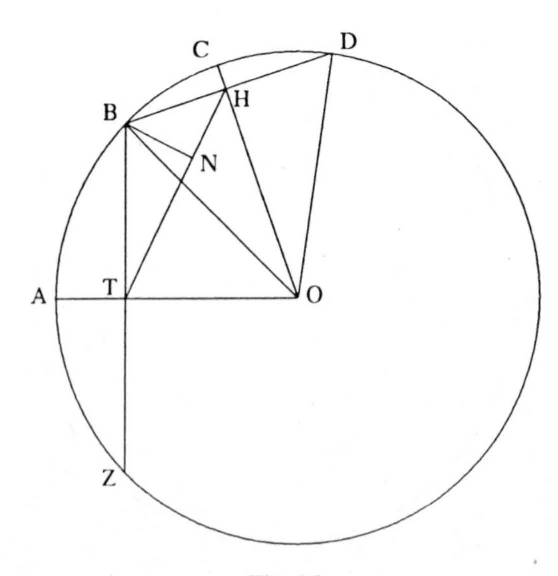

Fig. 5.9

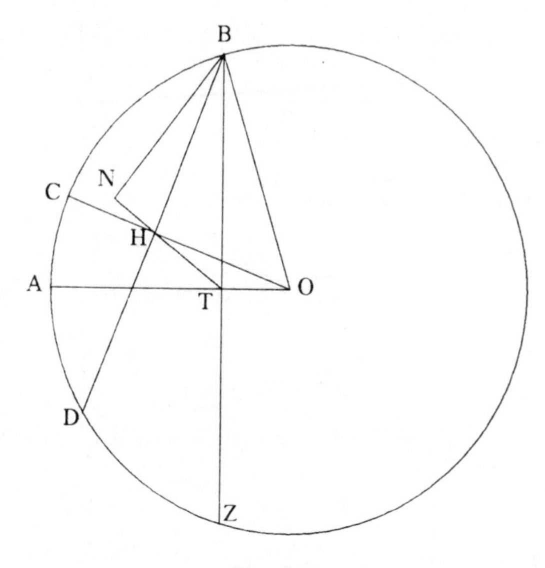

Fig. 5.10

nometric functions "expressed in sixtieths" does not mean the radius is 60 but only that his tables use sexagesimal fractions.)

Let there be two arcs AB, BC of a circle ABCD, and suppose we know the sine of each arc. I say that the sine of their sum, as well as that of their difference, is known. Join the three points A, B, C to the center O and from B drop BT, BH, perpendicular to the radii OA, OC, respectively, and then draw HT. In addition, prolong BH, BT to meet the circle at D, Z respectively. Since radii that are perpendicular to chords bisect these chords, BH = HD and BT = TZ. Thus triangles BHT, BDZ are similar and so DZ = 2TH. In Fig. 5.9, $\widehat{ZBD} = 2\widehat{AC}$ because $\widehat{ZB} = 2\widehat{AB}$ and $\widehat{BD} = 2\widehat{BC}$, and in Fig. 5.10, $\widehat{DZ} = 2\widehat{AC}$, for the same reasons and the fact that $\widehat{AB} - \widehat{BC} = \widehat{AC}$. Thus $TH = \frac{1}{2}DZ = \frac{1}{2}\,\mathrm{Crd}(2\widehat{AC})$, so $TH = \mathrm{Sin}\,\widehat{AC}$. To complete the preliminaries draw BN perpendicular to TH.

The key to Abu l-Wafā''s proof is his observation that since the angles BTO, BHO are right the four points B, T, H, O lie on the circumference of a circle of diameter BO, by Euclid's *Elements* III,31. In both cases (Figs. 5.9 and 5.10) the angles BHT, BOT subtend the same chord. In the first case (Fig. 5.9) they are on the same side of the chord BT so they are equal. In the second case (Fig. 5.10) they are on opposite sides of the chord BT, so they are supplementary. In this case the supplement of $\measuredangle$BHT, namely $\measuredangle$BHN, equals $\measuredangle$BOT. In both cases, then, the triangles BHN, BOT are right triangles with equal angles at H, O so the triangles are similar and thus BH/HN = BO/OT. Since $BH = \sin(\widehat{BC})$, $OT = \cos\widehat{AB}$ and BO = 1 it follows that $HN = \sin\widehat{BC}\cdot\cos\widehat{AB}$. In addition, the triangles BNT, BHO are similar because the angles N and H are right while the angles T and O are equal, being constructed on the chord BH in the circle through B, H, T and O. Thus

$$\frac{BT}{TN} = \frac{BO}{OH}, \quad \text{where} \quad BT = \sin(\widehat{AB}), \quad BO = a \quad \text{and} \quad OH = \cos \widehat{BC}.$$

Thus $TN = \sin(\widehat{AB}) \cdot \cos(\widehat{BC})$. Finally, in the case of Fig. 5.9,

$$\begin{aligned}\sin(\widehat{AB} + \widehat{BC}) &= \sin(\widehat{AC}) = TH = TN + NH \\ &= \sin(\widehat{AB}) \cdot \cos(\widehat{AC}) + \sin(\widehat{BC}) \cdot \cos(\widehat{AB}).\end{aligned}$$

In the case of Fig. 5.10,

$$\begin{aligned}\sin(\widehat{AB} - \widehat{BC}) &= \sin(\widehat{AC}) = TH = TN - NH \\ &= \sin(\widehat{AB}) \cdot \cos(\widehat{BC}) - \sin(\widehat{BC}) \cdot \cos(\widehat{AB}).\end{aligned}$$

§4. Naṣīr al-Dīn's Proof of the Sine Law

Naṣīr al-Dīn introduces the Sine Law for plane triangles to provide a basic tool for solving them, and in this section we shall see how he proves the law and how he applies it to find unknown parts of triangles from known ones.

The Sine Law. If ABC is any triangle then $c/b = \text{Sin } C/\text{Sin } B$.

Figure 5.11 illustrates the case when one of the angles B or C is obtuse, and Fig. 5.12 the case when neither B nor C is obtuse, so that one of them is acute. In either case prolong CA to D and BA to T so each is 60 units long and, with centers B, C, draw the circular arcs TH and DE. If we now drop perpendiculars TK and DF to the base BC, extended if necessary, then TK = Sin B and DF = Sin C. (In the case of Fig. 5.11 both of these statements are obvious, but in the case of Fig. 5.10 the reader must remember that

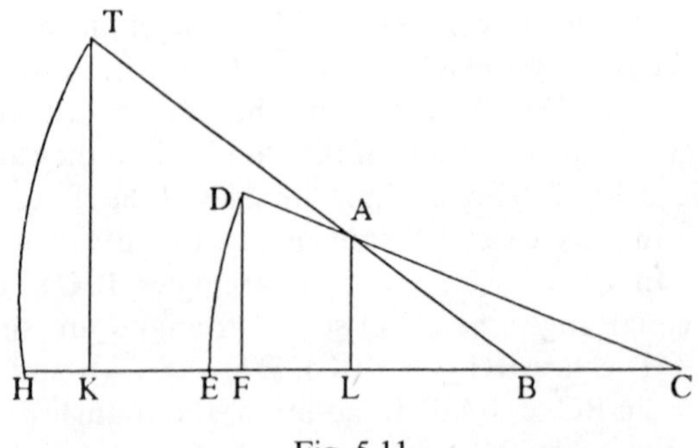

Fig. 5.11

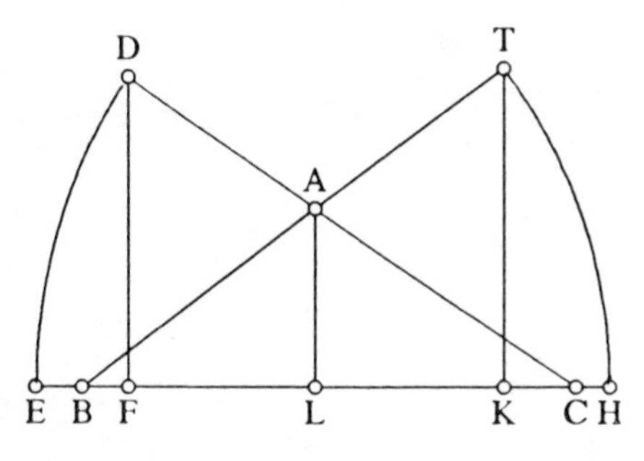

Fig. 5.12

Sin(∡B) = Sin(180° − ∡B).) Now draw AL perpendicular to BC. Since triangles ABL, TBK are similar AB/AL = TB/TK, and since triangles ACL and DCF are similar, AL/AC = DF/DC but DC = 60 = TB, so, if we multiply the left and right sides, respectively, of these two proportions, we obtain the proportion AB/AC = DF/TK. Therefore c/b = Sin C/Sin B, and this proves the Sine Theorem.

Since Naṣīr al-Dīn's sine function is simply 60 times the modern one, the above theorem holds for the modern function as well. We may re-write the theorem as $c/\sin C = b/\sin B = a/\sin A$, a form it is often given in today, and it may be most easily remembered as the statement that in a given triangle the ratio of any side to the sine of the opposite angle is constant.

Naṣīr al-Dīn uses this theorem to solve all possible triangles systematically, as follows.

Case 1. Two angles and one side known:

If two angles (A and B) are known then the other angle C = 180° − (A + B) is also known (Fig. 5.13), but at least one side must be given as well since a triangle cannot be determined from its angles alone. Since all angles are known, however, we may suppose, without loss of generality, that the known side is c. Then, by the Sine Theorem,

$$\frac{c}{b} = \frac{\text{Sin C}}{\text{Sin B}} \quad \text{and} \quad \frac{c}{a} = \frac{\text{Sin C}}{\text{Sin A}}.$$

In each proportion three out of the four terms are known and so the remaining terms, a and b, may be found.

Case 2. Angle and two sides known:

If only one angle is known then two sides must be known. If one of these is opposite the known angle then without loss of generality, the known are c, C and a. Then A is determined by c/a = Sin C/Sin A, and, since two angles are now known, we are back to the previous case, which we have already shown how to solve.

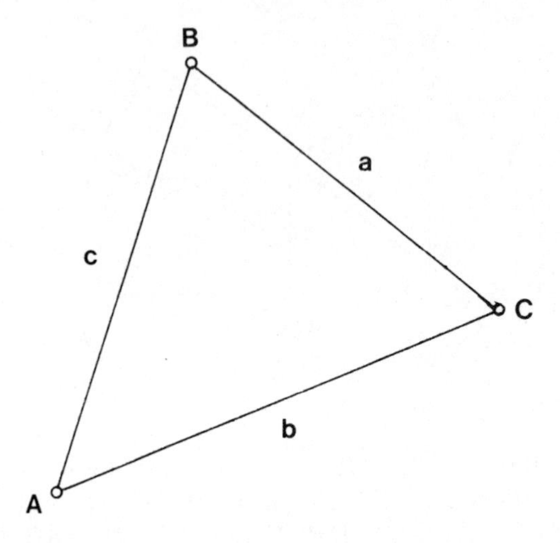

Fig. 5.13

If, on the other hand, neither side is opposite the known angle then, without loss of generality, the known are B, a, c (Fig. 5.14). In this case consider the perpendicular AE from A onto a. Then in the right-angle triangle BEA the side c and the angle B are known and the side AE may be found as described in the section on solving right triangles. Then $EB = \sqrt{BA^2 - AE^2}$ and $CE = a - EB$, so in the right triangle AEC two sides AE and EC are known and the remaining side AC $(=b)$ and the angle C may be calculated. Then $A = 180° - (B + C)$ and all parts of triangle ABC are determined.

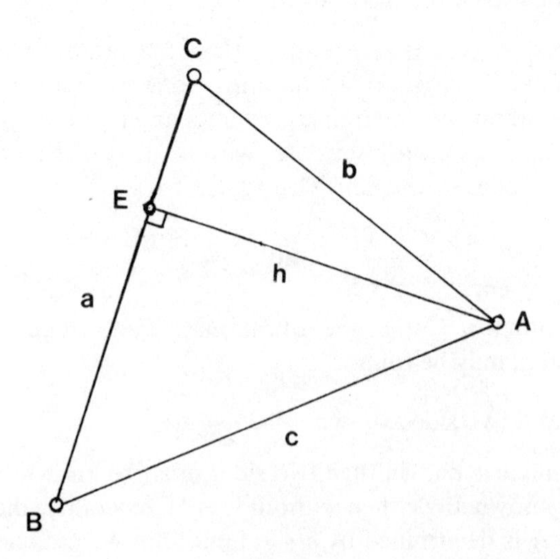

Fig. 5.14

Case 3. Three sides known:

If no angles are known then the three sides a, b and c are given. In this case (Fig. 5.14 also) Naṣīr al-Dīn says to calculate the perpendicular, h, from A onto a according to $\mathrm{BE} = (c^2 + a^2 - b^2)/2a$, and then $h = \sqrt{c^2 - \mathrm{BE}^2}$. He calls this "the usual rule" for calculating the perpendicular, and it may be found in Ptolemy's *Almagest*, Book VI,17. Since EC = BC − BE, in the two right triangles BEA, CEA the three sides of each are known. His solutions of right triangles show how to find the angles of such triangles, so B and C may be calculated as in that section and, from them, $\mathrm{A} = 180° - (\mathrm{B} + \mathrm{C})$.

§5. Al-Bīrūnī's Measurement of the Earth

An elegant application of elementary trigonometry was done by al-Bīrūnī when he was travelling with King Maḥmūd of Ghazna in what is now northwest India, but was in al-Bīrūnī's time known as al-Hind. In this section we shall follow his account of the method as he gives it in his work *On the Determination of the Coordinates of Cities.*

He begins with an account of a method Ptolemy gives in his *Geography*, which requires a geodetic survey to determine the distance along a great circle connecting two places of known latitude and longitude. Ptolemy's method is, in fact, a generalization of that of Eratosthenes, who measured along a meridian. Al-Bīrūnī introduces his method with his characteristic dry humor, "Here is another method for the determination of the circumference of the earth. It does not require walking in deserts."

Al-Bīrūnī states that the astronomer Sanad ibn ʿAlī was with the caliph al-Ma'mūn on one of his campaigns against the Byzantines and used this method when they came to a high mountain near the sea. Since the method assumes one knows how to determine the height of the mountain, al-Bīrūnī first explains how to do that. The problem is nontrivial since a mountain is not a pole and therefore we cannot easily measure the distance from us to the point within the mountain where the perpendicular from its summit hits ground level.

To measure the height of a mountain al-Bīrūnī first requires that we prepare a square board ABGD whose side AB is ruled into equal divisions and which has pegs at the corners B, G. Then, at D, we must set a ruler, ruled with the same divisions as the edge AB and free to rotate around D. It should be as long as the diagonal of the square. Set the apparatus as in Fig. 5.15 so that the board is perpendicular to the ground and the line of sight from G to B just touches the summit of the mountain. Fix the board there and let H be the foot of the perpendicular from D. Also, rotate the ruler around D until, looking along it, the mountain peak is sighted along the ruler's edge DT. Now since AD is parallel to EG, $\sphericalangle \mathrm{ADT} = \sphericalangle \mathrm{DEG}$, and therefore the right

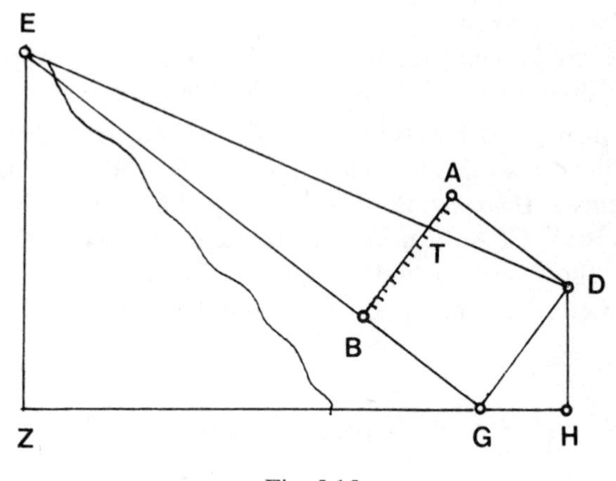

Fig. 5.15

triangles ADT and GED are similar. Thus TA:AD = DG:GE, and since of the four quantities in this proportion only GE is unknown we may solve for GE = AD · DG/TA. However, since both ∢EGZ + ∢DGH and ∢EGZ + ∢GEZ are equal to right angles it follows that ∢DGH = ∢GEZ, and thus the two right triangles DGH and GEZ are similar, so that GE:EZ = DG:GH. This means that we may solve for the single unknown EZ = GE · GH/DG, which is the desired height.

Adjusting the ruler through such a small angle as ADT is going to present problems, but al-Bīrūnī tells us he used the method to obtain reasonable results, so we shall take his word for it that the method is not totally impractical.

In fact, he says that "When I happened to be living in the fort of Nandana in the land of India, I observed from an adjacent high mountain standing west of the fort a large plain lying south of the mountain. It occurred to me I should examine this method there." He is referring to the method of finding the circumference of the earth, and it is the following, illustrated in Fig. 5.16.

Let KL be the radius of the earth and EL the height of the mountain. Let ABGD be a large ring whose edge is graduated in degrees and minutes, and let ZEH be a rotatable ruler, along which one can sight, which runs through E, the center of the ring. An astrolabe, which we describe in the next chapter, would be perfectly suitable for this, using the ruler and scale of degrees on the back of this instrument.

Now move the ruler from a horizontal position BED until you can see the horizon, at T, along it. The angle BEZ is called the dip angle, d. From L on the earth, imagine LO drawn so LO is tangent to the earth at L. By the law of sines applied to △(ELO)

$$\text{EL}:\text{LO} = \text{Sin(O)}:\text{Sin(E)} = \text{Sin}\, d:\text{Sin}(90° - d).$$

Since the two angles, as well as EL, the height of the mountain, are

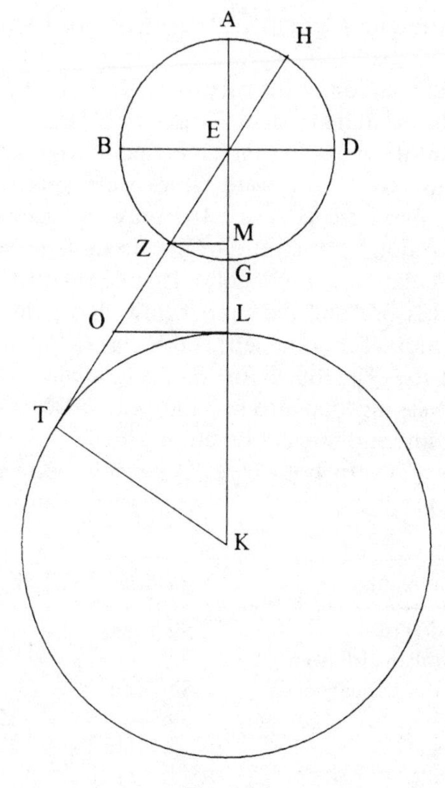

Fig. 5.16

known, we may determine LO; but, TO = LO, since both are tangents to a circle from a point O outside it. Also, since EL and LO are known, it follows from the Pythagorean theorem that $EO = \sqrt{EL^2 + LO^2}$ is known, and hence ET = EO + OT is known. Again, by the Law of Sines, ET : KT = Sin(d) : Sin(90° − d). Since KT, the radius of the earth, is the only unknown quantity in this proportion, we may solve for KT and so find the radius of the earth.

As al-Bīrūnī said, he tried the method on a mountain near Nandana in India where the height, EL, was 652;3,18 cubits and the dip angle was 34′. (Note the very small angles and the rather optimistically accurate height.) These give for the radius of the earth 12,803,337;2,9 cubits. Al-Bīrūnī takes the value $3\frac{1}{7}$ for π and arrives at the value 80,478,118;30,39 cubits for the circumference of the earth, which, upon division by 360 yields the value of 55;53,15 miles/degree on a meridian of the earth. This, he notes, is "very close" to the value of 56 miles/degree determined by a geodetic survey undertaken at the time of al-Ma'mūn. It doubtless gladdened al-Bīrūnī's heart to show that a simple mathematical argument combined with a measurement could do as well as two teams of surveyors tramping about in the desert.

§6. Trigonometric Tables: Calculation and Interpolation

The successful application of the rules that Naṣīr al-Dīn gives for finding the unknown parts of triangles from known parts depends not only on knowing the relevant theorems but on having good trigonometric tables and knowing how to use them. Accurate tables were necessary, not only to advance the study of such sciences as astronomy and geography but also to investigate such questions as the ratio of the circumference of a circle to its diameter. In the calculation of such tables, Islamic scientists went far beyond their ancient predecessors, and the chart below shows the increasing precision in the construction of trigonometric tables, as found in the *zīj*es of some of the major scientists. (The fourth line of the following table, for example, should be read as saying that around the year 1030 in his *zīj* al-Bīrūnī tabulated both the sine and tangent functions, the former in steps of 15′ and the latter in steps of 1°. In both cases the results were accurate to four sexagesimal places.)

Year	Individual	Function(s)	Intervals	Places
850	Ḥabash al-Ḥāsib	Sin, Tan	1°	3
900	Abū ʿAbdullah al-Battanī	Sin	$\frac{1}{2}$°	3
1000	Kūshyār ibn Labbān	Sin, Tan	1′, 1°	3
1030	al-Bīrūnī	Sin, Tan	15′, 1°	4
1440	Ulūgh Beg	Sin, Tan	1′	5

Consider what the calculation of a set of sine tables such as those of Ulūgh Beg involved. First, sixty entries must be done for each of 90 degrees, so a total of 5400 entries must be calculated. Moreover, the table will be only as accurate as the basic entries from which the others are calculated, for, from the value of Sin(1°) and from those of a few other basic Sines, the trigonometric formulae already discussed will yield a table of Sines for all integer values of n°. The half-angle formulae may then be used to fill in the intervals of $\frac{1}{2}$° or $\frac{1}{4}$°, and for divisions finer than this some sort of interpolation would have been used, which, of course, still demands computation. Bear in mind, also, that the tables often came provided with auxiliary columns, recording the increment or decrement from one line to the next to aid the user in performing interpolations, and you will understand the amount of labor required, in a pre-calculator era, to produce such a set of tables.

§7. Auxiliary Functions

The application of trigonometry to construct astronomical tables often requires the repeated calculation of the same combination of trigonometric functions, but for various values of the arguments. For example, in calculating tables of solutions to problems in spherical astronomy expressions which frequently occur are

$$\frac{\operatorname{Tan}\theta\cdot\operatorname{Sin}\varepsilon}{R}\quad\text{and}\quad\operatorname{arc\,Tan}\left(\frac{\operatorname{Tan}\varepsilon\cdot\operatorname{Sin}\theta}{R}\right),$$

where $\varepsilon = 23\frac{1}{2}°$ (approximately), R is the radius of the circle used to define the trigonometric functions, and θ runs over a certain number of degrees. Thus, in computing tables for use in astronomy, timekeeping and regulating the times of prayer Muslim astronomers noticed that, in the course of the computations, they were computing the same things over and over again. Only the value of the argument was varying.

It soon occurred to some of the astronomers to lessen the labor of computing tables of functions by computing a set of tables of those auxiliary functions that appeared often. As early as the mid-ninth century Ḥabash al-Ḥāsib tabulated, among other functions, the two given above, for $R = 60$. Later in the same century al-Faḍl al-Nayrīzī again tabulated the two functions above for $R = 150$, a parameter common in Indian trigonometry. Then, late in the tenth century, al-Bīrūnī's teacher, Prince Abū Naṣr, tabulated four such auxiliary functions with $R = 1$, perhaps as much to show the utility of this value of the radius as for any other reason. At about the same time, in Ibn Yūnus' works, we find auxiliary functions which are functions of two arguments, but in which one of them (the latitude) assumes only the value of the latitude of Cairo or, sometimes, of Baghdad. The next major step in the calculation of auxiliary tables was taken by the fourteenth century astronomer Muḥammad al-Khalīlī of Damascus who generalized Ibn Yūnus' tables and calculated the following two-argument functions:

1. $f(\varphi, \theta) = R\cdot\sin\theta/\operatorname{Cos}\varphi$, $\theta = 1°, \ldots, 90°$; $\varphi = 1°, \ldots, 55°$, 21;30° (the latitude of Mecca), 33;31° (the latitude of Damascus).
2. $g(\varphi, \theta) = \operatorname{Sin}\theta\cdot\operatorname{Tan}\varphi/R$, for θ, φ as above.
3. $G(x, y) = \operatorname{arc\,Cos}(xR/\operatorname{Cos} y)$, $x = 1, \ldots, 59°$, $y = 0°, 1°, \ldots, n(x)°$, where $n(x)$ is the largest integer such that $x\cdot R$ does not exceed $\operatorname{Cos}(n(x))$. (Recall that $\operatorname{Cos}\theta$ is a decreasing function.)

Al Khalīlī's tables, which contain over 13,000 entries, allow one to solve any of the fundamental problems of spherical astronomy for any latitude and so represent general solutions to these problems. In the following chapter we shall see that al-Khalīlī also provided such a solution in the case of the direction of Muslim prayer.

§8. Interpolation Procedures

Out of the mass of labor in computing the tables described in the previous sections two parts require particular mathematical insight: (1) Calculating Sin 1°; and (2) Framing rules to allow the user to interpolate from the tables.

Zījes contain a variety of ingenious methods of interpolation, and we shall discuss an example of second-order interpolation. Following that we shall study an interative method of al-Kāshī that allows the rapid calculation of Sin(1°), accurate to as many places as one wishes.

Linear Interpolation

Figure 5.17 shows part of the table of sines from al-Bīrūnī's *Mas'ūdic Canon* together with a transcription of the table from the alphabetic notation into our numerals. Al-Bīrūnī's sine function is the same as the modern one, since for each θ, $0 < \text{Sin}\,\theta < 1$. The only difference is the conceptual one that we view the sine as a function of a central angle, while al-Bīrūnī saw it as a quantity that depends on the arc. A further difference, not apparent from the table, is that al-Bīrūnī's function assumes only positive values for, in the *Canon*, he tells the reader that:

(1) When $90° < \theta < 180°$ then $\sin\theta = \sin(180° - \theta)$, a fact we have already seen Naṣīr al-Dīn use in his proof of the Sine Theorem.
(2) When $180° < \theta < 270°$ then $\sin\theta = \sin(\theta - 180°)$.
(3) When $270° < \theta < 360°$ then $\sin\theta = \sin(360° - \theta)$.

Thus al-Bīrūnī's sine function is the absolute value of our modern function, and since neither al-Bīrūnī nor any other Muslim astronomer had the conve-

الفضول			التعاديل				الجيوب				دقائق	درج
روابع	ثوالث	ثواني	روابع	ثوالث	ثواني	دقائق	روابع	ثوالث	ثواني	دقائق	عدد القسي	سطر
كيح	مب	ية	نب	مط	ب	ا	كح	مب	ية	•	ية	•
كه	مب	ية	م	مط	ب	ا	نو	كد	لا	•	ل	•
كب	مب	ية	كح	مط	ب	ا	كا	ز	مز	•	مه	•
يج	مب	ية	يب	مط	ب	ا	مج	مط	ب	ا	•	ا
يب	مب	ية	مح	مح	ب	ا	ا	لب	يج	ا	يه	ا
و	مب	ية	كد	مح	ب	ا	يج	يد	لد	ا	ل	ا
نح	ما	يه	نب	مز	ب	ا	يط	نو	مط	ا	مه	ا

Fig. 5.17(a)

Degrees	Minutes	Sines				Differences			Corrections			
Column of Arc number		Minutes	Seconds	Thirds	Fourths	Seconds	Thirds	Fourths	Minutes	Seconds	Thirds	Fourths
0	15	0	15	42	28	15	42	28	1	2	49	52
0	30	0	31	24	56	15	42	25	1	2	49	40
0	45	0	47	7	21	15	42	22	1	2	49	28
1	0	1	2	49	43	15	42	18	1	2	49	12
1	15	1	18	32	1	15	42	12	1	2	48	48
1	30	1	34	14	13	15	42	6	1	2	48	24
1	45	1	49	56	19	15	41	58	1	2	47	52

Fig. 5.17(b). Transcription of part of al-Bīrūnī's Sine Tables

nience of negative numbers, it was often necessary to break an argument up into several cases in order to specify in each case the direction in which a given line segment or arc should be taken.

In Fig. 5.17 there are four columns, headed "Degrees/Minutes", "The Sines", "The Corrections", "The Differences", explained as follows. The first entry in Col. 1 is 15′ and each succeeding entry is 15′ more than the previous entry, so Col. 1 lists arcs θ, by increments of 15′ from 15′ to 90°. Column 2 then gives, for a given arc θ, the value of $\sin\theta$, so that, for example, the third row of this column may be read as saying $\sin 45' =$;0,47,7,21. (The reader may want to think for a minute to discover what the next two columns contain.) Col. 3, headed "Differences", records for each θ the value $\sin(\theta + 15') - \sin\theta = D(\theta)$ and Col. 4 simply gives, opposite the value θ, the value $C(\theta) = 4D(\theta)$.

Exercise 5 makes it clear that the fourth column in al-Bīrūnī's tables facilitates linear interpolation, and such interpolation works very well for the sine function. (In fact, if the reader converts the sexagesimal value for $\sin(1°22')$ obtained from Exercise 5 to decimal values the result is 0.02385051, rounded to eight places, whereas the correct value, rounded to eight places, is the only slightly higher value .02385057.)

In general, linear interpolation works well over small intervals for functions whose rate of growth does not change too much over the interval. However, for a function such as the tangent function, which has a vertical asymptote at 90°, linear interpolation does not produce satisfactory results and more refined methods are necessary. Such methods were developed early in the history of trigonometry, in fact in India before the Islamic era.

Muslim scientists also needed interpolation for other purposes. For example, they needed ephemerides, sets of tables giving the positions of the sun, moon and planets at equally spaced intervals (e.g. of one day or five days) throughout the year, but to compute such a set of tables was a great deal of work, often involving the production of various auxiliary tables and further computations based on these. In order to lighten this labor, Muslim astronomers would use interpolation procedures in compiling the tables, just as we know they used them in applying the completed ephemerides to find the position of a planet at a time not entered in the tables. In the next chapter, when we survey some of the tables used in spherical astronomy, which could have anything from 30,000 to 250,000 entries, the reader will see other ways in which interpolation methods were used.

Ibn Yūnus' Second-Order Interpolation Scheme

For the present, however, we are going to study a method of interpolation that appears in *al-zīj al-Ḥākimī* composed by the Egyptian astronomer Abu l-Ḥasan ibn Yūnus, the son of an eminent Egyptian historian who became one of the great astronomers of the Islamic world. We do not know precisely when he was born, but since his father died in 958 Ibn Yūnus witnessed not only the conquest of Egypt by the Fatimid kings in 969 but also the foundation of the city of Cairo by the same dynasty, a dynasty which claimed descent from Fāṭima, daughter of the Prophet Muḥammad.

Ibn Yūnus counted as his patrons at least two of the Fatimid kings, al-ʿAzīz, who reigned for roughly 20 years until 996 and al-Ḥākim, a firm believer in astrology who considered himself to be God.

However, Ibn Yūnus yielded to no one in eccentricity. Al-Ḥākim himself told the following story of Ibn Yūnus' indifference to custom. It seems that Ibn Yūnus came into his presence one day carrying a pair of heavy shoes. He sat near al-Ḥākim for awhile, while the ruler eyed the shoes—objects which court etiquette required be left outside the throne-room. Finally, Ibn Yūnus kissed the ground, put the shoes on, and left. Another time Ibn Yūnus and a fellow-astronomer went up into the Muqaṭṭam hills outside of Cairo. They observed Venus for some time, and then, according to the biographer Ibn Khallikān, "Ibn Yūnus took off his cloak and turban. Then he put on a woman's red cloak and a red veil, and took out a lute. This he played on, with incense burning in front of him. It was a remarkable sight." (Transl. from King (1972).) In addition to his expertise on the lute Ibn Yūnus achieved renown as a poet, and several of his works were contained in anthologies.

Ibn Yūnus named his great set of astronomical tables *al-zīj al-Ḥākimī* after his patron al-Ḥākim, who probably appreciated Ibn Yūnus more for the reputed accuracy of his astrological predictions. The historian Ibn Abī Ḥajala tells the following story:

> Another example of his correct (astrological) predictions took place when al-Ḥākim had given him a house. He said, "Prince of the Believers, I desire that you give me another house." (Al-Ḥākim) asked why. He said, "Because water

will destroy it (the one I have) and everything in it." Al-Ḥākim gave him another and he moved out (of the first) the next morning. Three days later a mighty torrent came down on Cairo from the mountain and threw down palaces and houses—a frightening event such as had never been experienced before—and the above-mentioned house was among what was destroyed, as he had predicted (King, 1972).

The same source says Ibn Yūnus predicted the day of his death and, after he had locked himself in his house, he told his servant girl, "Iḥsān, I have locked what I shall never open." He then took some water and began washing the ink off his manuscripts, and finally, continually reciting the Quranic verse "Say God is one", he died. This was in the year 1009.

The story of Ibn Yūnus washing the ink off his manuscripts is consistent with another story that after his death his son sold all his books, by the pound, in the Cairo soap market. After the washing there may have been little left to do with them but sell the remainder for scrap.

Ibn Yūnus intended that the interpolation procedure he described be used in trigonometric tables calculated for intervals of $30'$ (see Plate 5.1). (However, the extract from his Sine tables (shown in Plate 5.1) shows that he tabulated the Sine for intervals of $1°$.) In the following account we shall consider an arc $\theta + k'$, where θ is an integer number of degrees and $0' < k' < 60'$, and LSin will denote the value of the Sine obtained by linear interpolation. The procedure, then, is the following:

(1) Use linear interpolation between successive degrees to find $\mathrm{LSin}(\theta + k')$ and $\mathrm{LSin}(\theta + 30')$.
(2) From the table find $\mathrm{Sin}(\theta + 30')$.
(3) Define the "base for the interpolation" to be $4(\mathrm{Sin}(\theta + 30') - \mathrm{LSin}(\theta + 30')) = \mathrm{B}$. (Ibn Yūnus has already observed that Sines found by linear interpolation are always less than the actual values.)
(4) Calculate $\mathrm{B} \cdot k' \cdot (60' - k')$, and note $\mathrm{B} \cdot k' \cdot (60' - k') = \mathrm{B} \cdot k \cdot (60 - k)/3600$.
(5) For the value of $\mathrm{Sin}(\theta + k')$ take $\mathrm{LSin}(\theta + k') + \mathrm{B} \cdot k \cdot (60' - k')$.

It seems that the discoverer of this method had the idea of beginning with linear interpolation over a given $1°$-interval and then correcting by an amount that would, at the midpoint of the interval, bring the computed value to the true value. Perhaps it was experience with tables and interpolation, rather than geometrical arguments, that taught astronomers that not only is LSin θ less than Sin θ but that, over a given interval, the difference between these two would be greatest (or nearly so) at the midpoint. The problem, then, is to find a function, f, defined from $0'$ to $60'$ so that $f(0') = f(60') = 0$ and $f(30')$ is the greatest value, for then

$$\mathrm{LSin}(\theta + k') + f(k') \cdot \frac{\mathrm{Sin}(\theta + 30') - \mathrm{LSin}(\theta + 30')}{f(30)}$$

will be a good rule for interpolation. It produces the value $\mathrm{Sin}(\theta + 30')$ when $k' = 30'$, and elsewhere a value that lies between the value obtained by linear

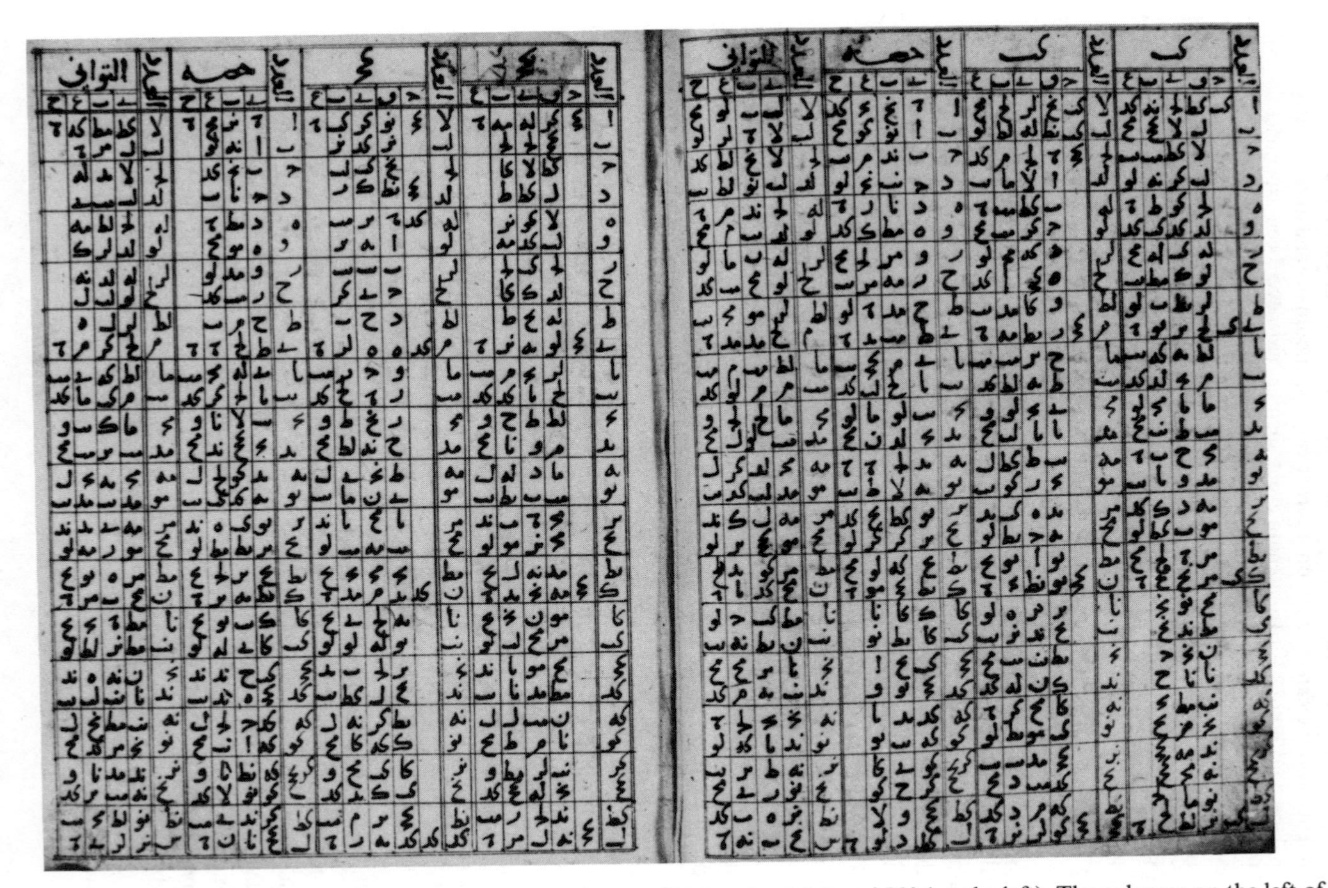

Plate 5.1. Since tables attributed to Ibn Yūnus. Extract for arguments 22° (on the right) and 23° (on the left). The columns on the left of each page are for interpolation. The horizontal argument is the number of degrees (22° and 23° in this case) and the vertical argument is minutes ([illegible] 20′ and 31′, 60′). (Taken from MS Berlin Staatsbibliothek, Ahlwardt 5752 fols. 13^v–14^r (Lbg 1038).)

interpolation and the true value. In fact, one may take $f(x) = x \cdot (60 - x)$, and any competent mathematician from the time of Euclid onward would have realized that the value of the product $x \cdot (a - x)$ is greatest when the two factors x and $a - x$ are equal. When $a = 60$, this implies that $x = 30$, which yields the maximum value of f as 900″.

Ibn Yūnus makes no attempt to show that this rule produces better approximations than linear interpolation, and it is typical of mathematics in the ancient and medieval worlds that no attempt was made to axiomatize these numerical methods or to give proofs of their validity. For the scientists of Ibn Yūnus' time these rules were only procedures, doubtless arrived at by plausible reasoning, but not proved, which the practitioner would find worked. See Plate 5.1 for an extract from the Sine tables composed by Ibn Yūnus.

§9. Al-Kāshī's Approximation to Sin(1°)

Interpolation procedures such as the one discussed in the previous section provide but one example of mathematical practice leading theory. Another example is featured in this section, an iterative solution to a third degree equation.

When Ptolemy approximated Crd(1°) in his *Almagest* he made use of the inequality:

$$\tfrac{2}{3}\,\mathrm{Crd}(\tfrac{3}{2}^{\circ}) < \mathrm{Crd}(1^{\circ}) < \tfrac{4}{3}\,\mathrm{Crd}(\tfrac{3}{4}^{\circ}),$$

which gave him an approximation to Crd(1°) correct to two sexagesimal places, since the two extreme terms of the inequality both begin 1;2,50. However, the method has an inherent limitation in that, whatever be the bounds one uses on either side, they will agree only to a certain number of places (since they are not equal), and there is no possibility of getting greater accuracy without finding new bounds.

Refinements of Ptolemy's method were used by mathematicians of the Islamic period to find an approximation of Sin(1°), but it was Jamshīd al-Kāshī who, in Samarqand early in the fifteenth century, discovered a method which will provide arbitrarily close approximations to Sin(1°), a method based on two relationships:

$$\mathrm{Sin}(3\theta) = 3\,\mathrm{Sin}(\theta) - ;0,4(\mathrm{Sin}\,\theta)^3, \tag{1}$$

which, when $\theta = 1^{\circ}$, becomes

$$\mathrm{Sin}(3^{\circ}) = 3\,\mathrm{Sin}(1^{\circ}) - ;0,4(\mathrm{Sin}\,1^{\circ})^3,$$

and on

$$\mathrm{Sin}(3^{\circ}) = 3;8,24,33,59,34,28,15, \tag{2}$$

which is exact as far as it goes. This value for Sin(3°) can be determined as exactly as necessary because Euclidean procedures allow one to find both Sin(72°) and Sin(60°) from the constructions of the sides of a regular pentagon and an equilateral triangle in a given circle. These Euclidean constructions, when translated into algebraic equations, demand nothing more than the solution of first or second degree equations. The roots of these may be expressed in terms of, at worst, square roots, which may be approximated to any desired accuracy. Then the formula for the sine of the difference of two arcs, which we have seen was known to Abu l-Wafā' in the tenth century, would yield Sin(12°) = Sin(72° − 60°) to any desired accuracy, and this, by repeated use of the half-angle formula, would yield in turn Sin(6°) and Sin(3°). When we substitute this value of Sin(3°) into the equation and write x for Sin(1°) we obtain, after a bit of arithmetic, the fundamental relation:

$$x = \frac{x^3 + 47{,}6;8{,}29{,}53{,}37{,}3{,}45}{45{,}0}$$

a cubic equation one of whose roots is Sin(1°).

Al-Kāshī knows this equation has a root near 1, so we may write the root as $1;a,b,c,\ldots$ where $a,b,c,\ldots$ are the successive "sexagesimal places" of the root. Then substituting this value for x we obtain

$$1;a,b,c,\ldots = \frac{(1;a,b,c,\ldots)^3 + 47{,}6;8{,}29{,}53{,}37{,}3{,}45}{45{,}0}.$$

So, if we subtract 1 from each side,

$$;a,b,c,\ldots = \frac{(1;a,b,c,\ldots)^3 + 47{,}6;8{,}29,\ldots}{45{,}0} - 1$$

which simplifies to

$$;a,b,c,\ldots = \frac{(1;a,b,c,\ldots)^3 + 2{,}6;8{,}29,\ldots}{45{,}0}.$$

Place-by-place the two sides are equal, so in particular the first sexagesimal place of the right-hand side must be a. However, because 45,0 is so big and the root (and hence its cube) is near 1 the first place of the right-hand side does not depend on the value of a, To convince ourselves of this we can evaluate the right-hand side at 1;59 (or even 2) instead of at 1, and when we do we find that

$$\frac{2^3 + 2{,}6;8{,}29,\ldots}{45{,}0} = ;2,(58 \text{ or } 59),\ldots.$$

Thus to find out what a is we need only evaluate

$$\frac{(1)^3 + 2{,}6;8{,}29,\ldots}{45{,}0} = ;2,(49 \text{ or } 50),\ldots$$

so that $a = 2$.

Then again,

$$1;2,b,c,\ldots = \frac{(1;2,b,c,\ldots)^3 + 47,6;8,29,53,37,3,45}{45,0},$$

and al-Kāshī now uses the fact that the second digit on the right will not depend on b, because of the size of the divisor 45,0. (One can check this by evaluating the right-hand side with $x = 1;2$ and $x = 1;3$ to get 1;2,49,39, ... and 1;2,49,43, ... respectively.) Clearly, then, we may set $x = 1;2$ and obtain $b = 49$.

If, now, we write $f(x) = (x^3 + 47,6;8,29,\ldots)/45,0$ then we may state al-Kāshī's idea as follows: Since $f(x)$ increases so slowly near 1 the value of the nth digit of $f(x)$ does not depend on the value of the nth digit of x but only on the value of the first $n - 1$ digits of x. We have seen that, at least for the first two digits, the idea works, but will it always work?

Al-Kāshī does not address this question but continues computing to the ninth sexagesimal place (60^{-9}) and obtains the result $\text{Sin}(1°) =$ 1;2,49,43,11, ... 17, and one may check that for this value of x, $f(x) = x$, very nearly. Thus, al-Kāshī discovered a method for approximating Sin(1°) which will produce a value as near the true value as one wishes.

Methods such as the one al-Kāshī used are called *iterative* methods, which means that one begins with certain data and an approximation (generally one that is fairly crude but at least near) to the correct answer. One then uses the data and the initial approximation in a given procedure to arrive at a number. This number is then taken as the new approximation, and it, together with the data, is then put into the same procedure for a second round of computation. This computation produces another approximation which, together with the data, is again put into the procedure, etc. If the procedure is effective, the successive results will approach nearer and nearer to one value, which will be the value that solves the problem. In this case we say that the procedure converges and the algorithm, or procedure, is an effective one.

In fact, in al-Kāshī's algorithm the results of the successive approximations do approach the value of Sin(1°). The method is taught today in courses on numerical analysis under the name of "Fixed-point iteration", where it is proved that the procedure al-Kāshī used to find a solution A to the equation $x = f(x)$ will converge provided the curve $y = f(x)$ is a smooth one and the initial approximation is chosen in a neighborhood of A where the tangent line to the curve has a slope of absolute value less than 1.

Although the proof of the convergence of the algorithm does not go beyond the mean-value theorem of the differential calculus there is no evidence that al-Kāshī concerned himself with the proof of the theorem. He was working on the interface of the exact sciences and mathematics, the region that has been responsible historically for so much of the growth in mathematics, and his concern was to find methods which would provide solutions to problems important in astronomy. This he did, and, as we have seen, he did it well.

Having said this, we should also add that we do not yet know exactly how much, or how little, argument al-Kāshī gave to support his procedure. Until recently it was known only from a marginal note in a manuscript made by Miram Chelebi, the grandson of one of al-Kāshī's colleagues—and sometimes rival—at Samarqand. The method is described in the briefest possible way, and over the last century scholars have tried to interpret exactly what Chelebi was saying. It is now known that a manuscript discovered in Iran contains a treatise by al-Kāshī on finding Sin(1°), so we may one day know how, if at all, al-Kāshī imagined his procedure could be justified.

We have seen in the preceding sections that Muslim mathematicians organized trigonometry into a systematic discipline, whose theory rested on a full complement of six functions and a variety of powerful results, such as the Law of Sines and certain key trigonometric identities. In addition, the application of this theory was made possible by extensive, highly accurate trigonometric tables, including tables of auxiliary functions, a variety of methods for interpolation—a technique that has been nicely characterized as "reading between the lines", and iteration. Thus trigonometry takes its place, alongside algebra and arithmetic, as part of our heritage from Islamic mathematics.

Exercises

1. Discover the value of π implicit in the ancient Indian value of 3438′ for the radius of a circle, where, as we said in the text, the minute is the same length as a minute of arc on the circle.
2. Relative to Naṣīr al-Dīn's discussion of solving a right triangle when one acute angle and two sides are known, show that if the sides are a, b, g, and if b is taken to be $120 \cdot u$ then

$$a \text{ must be taken to be } a \cdot 120 \cdot u/b$$

and

$$g \text{ must be taken to be } g \cdot 120 \cdot u/b.$$

3. Use the value for $\mathrm{Crd}(\frac{1}{2}^\circ)$ given in Ptolemy's table of chords to estimate the value of π. How accurate is this estimate?
4. Use Naṣīr al-Dīn's explanation of how to solve right triangles, together with Ptolemy's table of chords, to solve the 3–4–5 right triangle.
5. Use al-Bīrūnī's sine table to calculate $\sin(1^\circ 22')$ from the formula

$$\sin(1^\circ 22') = \sin(1^\circ 15') + (;7)\mathrm{C}(1^\circ 15').$$

In general show that for any θ, $0 < \theta < 90^\circ$, If θ' is the largest multiple of 15′ that is less than θ then

$$\sin(\theta) = \sin(\theta') + (\theta - \theta')\mathrm{C}(\theta).$$

(Here, $\theta - \theta'$ must be read not as "minutes" but as sixtieths of the parts that make up the radius.)

6. Use Ibn Yūnus' procedure to calculate a value for Sin 1°22′, given the tabular values Sin 1° = 1;1.49,45, Sin 1°30′ = 1;34,14.13 and Sin 2° = 2;5,38.17. Show that your answer lies between the value produced by linear interpolation and the true value. Show also that for any value of the argument θ the value of Sin θ produced by Ibn Yūnus' procedure lies between that produced by linear interpolation and the true value.

7. Derive the final form of al-Kāshī's cubic equation as we have given it in the text from the form:

$$\mathrm{Sin}(3^\circ) = 3\ \mathrm{Sin}(1^\circ) - ;0,4\ (\mathrm{Sin}\ 1^\circ)^3,$$

(*Hint*: It helps to use the fact that (;0,4)/3 = ;0,1,20.)

8. If $x^2 = 2$ show that $x = (x + 2/x)/2$. Use al-Kāshī's method and the initial approximation $x = 1$ to calculate three more approximations to $\sqrt{2}$. Heron of Alexandria, who wrote in Greek ca. A.D. 60, recommends this method for approximating square roots in his *Metrica*. (Dr. C. Anagnostakis of New Haven, Conn. pointed out to me that this procedure is an example of al-Kāshī's algorithm.)

9. Let $g(x) = x^3 + 4x^2 - 10$, and show that the equation $g(x) = 0$ has a root in the interval (1,2). Now use algebra to show that this root is also a root of the equation $x = f(x)$, where $f(x) = \sqrt{10/(x+4)}$. Finally, starting with an initial approximation $x = 1$ to the root of this last equation, use al-Kāshī's method to find the successive decimal places of the root. Conclude that, to three places, the value of the root is 1.365.

10. The tenth-century writer Abū Ṣaqr al-Qabīṣī, in the context of arguing that the height of any known mountain is negligible compared to the radius of the earth, gives the following method of finding the height of a mountain. Justify it.

 Let BGD be the surface of the earth and AB the height of the mountain (Fig. 5.18). From the two points G,D, whose distance apart is assumed known, one measures the angles AGB and ADB. Then

$$\mathrm{AB} = \frac{\mathrm{GD} \cdot \mathrm{Sin(D)}}{\mathrm{Sin}(90^\circ - \mathrm{D}) - \mathrm{Sin}(90^\circ - \mathrm{G})\ \mathrm{Sin(D)/Sin(G)}}.$$

 Since all quantities on the right are known we can calculate AB.

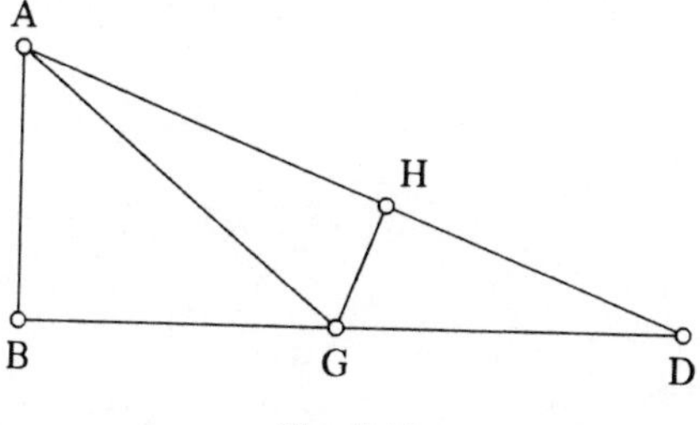

Fig. 5.18

Bibliography

Aaboe, A. "Al-Kāshī's Iteration Method for Sin(1°)", *Scripta Mathematica* **20** (1954), 24–29.

Hamadanizadeh, J. "A Survey of Medieval Islamic Interpolation Schemes". In: *Festschrift: A Volume of Studies of the History of Science in the Near East, Dedicated to E. S. Kennedy* (D. A. King and G. Saliba, eds.). New York: New York Academy of Sciences (to appear).

Kennedy, E. S. "The History of Trigonometry: An Overview". In: *Historical Topics for the Mathematics Classroom. Thirty-first Yearbook*. Washington, D.C: National Council of Teachers of Mathematics, 1969. Reprinted in E. S. Kennedy *et al. Studies in the Islamic Exact Sciences*. Beirut; American University of Beirut, 1983.

King, D. A. *The Astronomical Works of Ibn Yūnus*. (Unpublished Yale Ph.D. thesis.) New Haven: Yale University, 1972.

King, D. A. "Al-Khalīlī's Auxiliary Tables for Solving Problems of Spherical Astronomy", *Journal for the History of Astronomy* **4** (1973), 99–110.

al-Ṭūsī, Naṣīr al-Dīn (transl. by C. Caratheodory). *Traité du Quadrilatère*. Constantinople: 1891.

Chapter 6

Spherics in the Islamic World

§1. The Ancient Background

The problems of spherics ask for the calculation of the sizes of circular arcs or angles on the surface of a sphere. In applications, the sphere was either the celestial sphere or the earth, the former being a sphere that was thought to contain the fixed stars and to have such a large radius that, in relation to it, the earth was no more than a dot. However, the radius was finite and, for mathematical purposes, could be taken as the unit.

In the theoretical treatment of the geometry of the surface of the sphere the analogues of the straight lines on a plane are the *great circles*, which are the intersections of the surface of the sphere with any plane through its center. Also important are *parallel circles*, which are formed by the intersection of the surface of the sphere with planes that do not pass through its center.

In the case of the terrestrial sphere the important great circles are the equator and meridians, and the parallels of latitude are parallel circles. In the case of the celestial sphere some important great circles are the celestial equator, the ecliptic and the horizon, which we shall define and discuss in the next section. It is these interpretations of great circles and parallel circles that give the subject of spherics its utility both in astronomy and in mathematical geography.

It was the Greeks who first investigated the geometry of the surface of the sphere, and surviving treatises of Autolykos, who was probably a contemporary of Euclid in the fourth century B.C., show that the following Basic Facts were known long ago (Fig. 6.1(a)–(c)).

1. Any two great circles of a sphere bisect each other.
2. Given any two diametrically opposite points on the sphere, consider all great circles joining these two points. Then there is a unique great circle lying on the plane perpendicular to all those great circles. Conversely,

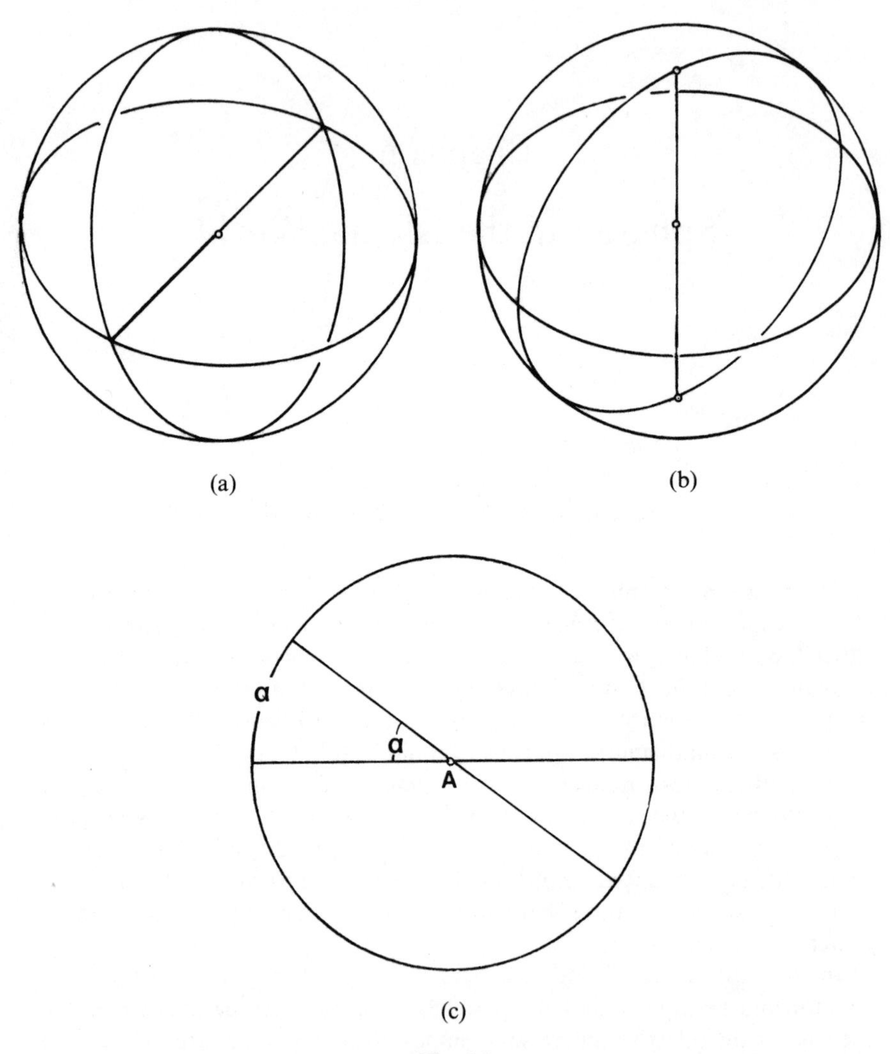

Fig. 6.1

given any great circle there are two diametrically opposite points called its poles such that any circle through these poles is perpendicular to the given circle.

3. Since any two given great circles of a sphere bisect each other it follows that they intersect at diametrically opposite points A, B. Then (2) guarantees that there is a unique great circle with A, B as poles. The smaller arc that is cut off from this circle by the two given circles measures the smaller angle α between these circles.

Menelaos, who conducted astronomical observations in Rome and lived a few decades before Ptolemy, is the first writer we know of to mention spher-

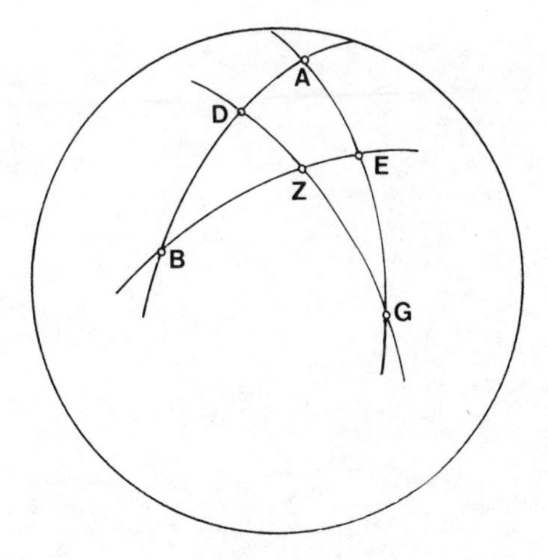

Fig. 6.2

ical triangles. In his work *Spherica* a spherical triangle is defined as "the area enclosed by the arcs of (three) great circles on a sphere, each arc being less than a semicircle", and it is in Book III of the *Spherica* that we find a theorem that is not only the first theorem of spherical trigonometry but, for the Greek writers, the only theorem of that science. It has become known as Menelaos' Theorem, and may be stated in a slightly modernized form as follows (Fig. 6.2):

> Let $\overset{\frown}{AB}$ and $\overset{\frown}{AG}$ be two arcs of great circles on the sphere, and let two other arcs $\overset{\frown}{GD}$ and $\overset{\frown}{BE}$ meet within the angle the first two arcs enclose, say at Z. In addition, let all four arcs be less than semicircles. Then
>
> $$\mathrm{Crd}(2\overset{\frown}{GA}):\mathrm{Crd}(2\overset{\frown}{EA}) = [\mathrm{Crd}(2\overset{\frown}{GD}):\mathrm{Crd}(2\overset{\frown}{ZD})]\cdot[\mathrm{Crd}(2\overset{\frown}{ZB}:\mathrm{Crd}(2\overset{\frown}{BE})].$$

As the reader may suspect, this is but one of many cases of the theorem. Ptolemy states one more, and medieval Islamic writers obtained 72 cases in all of what they called the "figure of the complete quadrilateral". For our purposes it suffices to note that it often requires ingenuity in completing a few given arcs to obtain a configuration to which one could apply the theorem, while a single triangle would be much easier to find. (For example, compare Exercises 5, 11 and 13 of this chapter.)

In addition to Menelaos' Theorem, other methods for finding arcs or angles on the sphere were also of some advantage, either in getting reasonable approximations easily or in demonstrating basic facts to the beginning student. One such method is simply to construct a good model of a sphere and then engrave on it the important great or parallel circles as well as the positions of the important stars (in the case of the celestial sphere) or the major geographical features (in the case of the earth). Then to solve problems of spherics one could simply use a substance such as colored wax or

Plate 6.1. An illustration of an armillary sphere taken from MS Vatican Borg. ar. 817, fol. 1ʳ. Reproduced courtesy of the Vatican Apostolic Library.

chalk to mark on any additional arcs or angles needed and then measure the ones desired.

Increasingly sophisticated variations on the above practical approach were described by a series of writers. For example, in the ninth century, Qusṭā ibn Lūqā wrote a treatise *On the Sphere with a Frame*, and in the twelfth century ʿAbd al-Raḥmān al-Khāzinī described an automated device of this sort in a treatise titled *On the Sphere that Moves by Itself*. (This treatise is described in *Lorch*.) Such devices were known not only to the Muslim authors but in the ancient world as well, and al-Khāzinī's treatise belongs to the tradition of powered models that can be traced back to Archimedes' moving model of the sun, moon and planets rotating around the earth. A simpler version of the same idea was the solid sphere surrounded by a framework of graduated rings, each corresponding to an important circle. Such an instrument is described in Ptolemy's *Almagest*, where it is called "an astrolabe", but we know it as an armillary sphere (from the Latin *armilla* = bracelet, for so the rings around the sphere would appear). Plate 6.1 shows an illustration of an armillary sphere from a manuscript in the Vatican.

§2. Important Circles on the Celestial Sphere

Since in much of the following we shall be concerned with the applications of spherics to the celestial sphere we shall here introduce the reader to some of the most important circles and angles on that sphere, and we shall begin with the one that is most apparent to the reader. In our explanation we shall use the capitalized word "Star" to denote a star, the sun, the moon, or any of the five naked-eye planets when the discussion applies equally to all of these objects.

When one looks around on a wide plain he sees all around him the line bounding sky and earth. Anything in the heavens that is above this line is visible and anything below it is invisible, for which reason the ancient Greeks called it *horizōn*, that is the "bounding" or "defining" circle, and from this we take our phrase "horizon circle". The points on the celestial sphere which are directly over our head and directly underneath our feet are the poles of this circle. The modern names for these poles, *zenith* and *nadir*, respectively, come from the Arabic words *samt*, "direction" (of the head), and *nāẓir* "opposite" (the feet). The great circles joining these two points are called *altitude circles*, and the one of these that passes through the north and south points of the horizon is called the *meridian* of our locality. If the reader imagines a Star and the altitude circle that passes through it he will easily see that the smaller arc of that circle between the Star and the horizon, measured in degrees, is a reasonable measure of the *height* of the Star above the horizon, hence the name *altitude* circle. Again, the smaller angle between the altitude circle of a Star and the meridian measures how many degrees one must turn from the N–S line to see the Star, and this angle is called the *azimuth* of the Star, again from the Arabic word *samt*, "direction" (of the Star).

One of the most striking daily phenomena in the heavens is the rising and setting of sun, moon and stars. The fixed point around which they appear to rotate each day is called a (celestial) pole, north or south depending on whether one is in the northern or southern hemisphere, for there are two poles, of which only one is visible to a given observer. (Note that stars sufficiently near the visible pole never set.) The circle a Star makes as it rotates around the pole in the course of 24 hours is called its *day-circle*, and that part of its day-circle that is visible is called, for obvious reasons, its *arc of visibility*.

The great circle perpendicular to all the great circles through the north and south poles is called the (celestial) equator. If one imagines these poles and circles somehow visible in the heavens then the equator would be seen just rotating into itself, going around the earth every 24 hours, but the great circles joining the poles and perpendicular to the equator would be seen actually rotating around the heavens, always passing through the fixed poles. Any Star lies on one of these circles and the smaller arc of the circle contained between some Star and the equator is called the *declination* (symbolized by "δ") of the Star, which may be thought of as measuring the height of the Star relative to the equator. Also, the angle between the circle passing through the Star and the meridian is called the *hour angle* of the Star, and is useful in telling time. For example, when the Star is the sun the hour angle measures the time until noon, using the conversion $15° = 1$ hour.

Finally, if one watches the skies a bit before sunrise he will see some star rise above the eastern horizon just before it becomes too light to see the stars any more. In other words, the sun is near that star. After another week or so he will again see a star rise just before sunrise, but it will not be the same star as before, which will have risen some time earlier, so it appears that the sun has moved relative to the stars, and it is now near another star. If one watches during the course of a year he will see the sun go completely around the heavens, at the rate of about a degree each day, and return to the same star. In fact, the sun appears to follow a great circle around the heavens, and this circle is called the ecliptic (from the Greek *ekleipein*, which means "to eclipse"). If one takes a narrow band, say 5° wide on either side of the ecliptic, then not only the sun, but the moon and the five naked-eye planets all appear to move relative to the stars within this band. When the moon is on the ecliptic there is a possibility of an eclipse, which is why the Greeks chose the name they did.

Since the ecliptic and equator are two great circles they intersect at diametrically opposite points, where they form an angle of approximately $23\frac{1}{2}°$, which is called the *obliquity of the ecliptic* and is written with the Greek letter

ε. The points of intersection are called the spring and fall *equinoctial points* (because when the sun is at these points day and night have equal length). One of these points, that for the spring, is taken as the 0° point for longitudes on the ecliptic, which are measured in the counterclockwise direction looking down on the ecliptic from the north.

The narrow band which surrounds the ecliptic on each side is called the *zodiac* and is divided into twelve segments, each 30° long. The names of the signs, beginning at the point of the spring equinox and running counterclockwise, are the following (reading across the page):

Aries	Taurus	Gemini
Cancer	Leo	Virgo
Libra	Scorpio	Sagittarius
Capricorn	Aquarius	Pisces.

According to this arrangement, the point which is 90° (counterclockwise) from the beginning of Aries, and is therefore the northernmost point on the ecliptic, is the beginning of the sign of Cancer. Thus the circle it makes as it turns during the day is called the Tropic (from the Greek *tropos*, "turning") of Cancer. Likewise, the southernmost point of the ecliptic (six signs away from the beginning of Cancer) is the beginning of Capricorn, so its turning creates the Tropic of Capricorn.

It seems to be one of mankind's enduring beliefs that the arrangement of the sun, moon and various planets within the zodiac as well as the placement of the zodiac relative to the horizon at the time of some event (the birth of an individual, the founding of a city, the beginning of a military campaign) influence for good or ill the outcome of the event. Thus there arose a set of motives for study of the movement of the heavenly bodies that included self-interest, in addition to such motives of a practical or scientific nature as the construction of calendars or reckoning time.

§3. The Rising Times of the Zodiacal Signs

A typical problem of spherics which makes use of both the equator and the ecliptic is finding the rising times of arcs on the ecliptic. Thus, Fig. 6.3 shows the celestial sphere at a time when the point on the ecliptic of longitude λ is rising above the eastern horizon of a locality. Imagine the sun to be at this point, so it too is just rising on some day of the year, and disregard the slow motion of the sun on the ecliptic, which, as we said, amounts to a little less than 1°/day. Then, at sunset of that day, the sun will still have longitude λ, but it will now be on the western horizon. Since any two great circles bisect

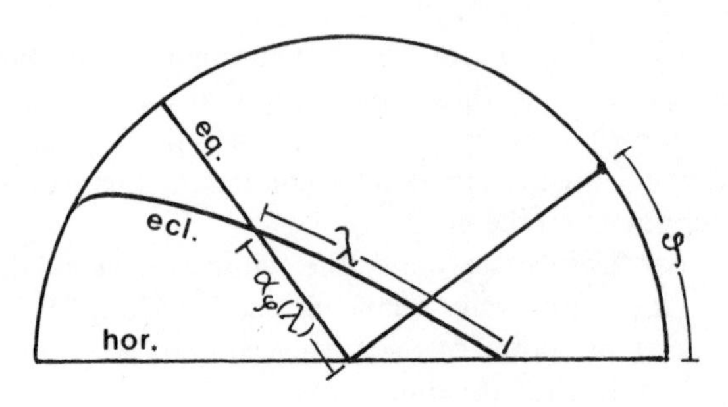

Fig. 6.3

each other, half of the ecliptic will be above the horizon at any time of the day, in particular at sunset, and this half of the ecliptic will have risen during the day as the point where the sun is travels across the sky. Hence, during each daylight period, 180° of the ecliptic rises over the horizon. Thus, if we can tell, for any arc of the ecliptic, how long it takes for it to rise over our horizon (called the *rising time* of that arc) we will be able to calculate how long daylight will last on any given day (provided we know where the sun is on that day, of course). In ancient and medieval geography the length of the longest day of the year at a given locality was one measure of the local latitude, and the length of daylight was also important in telling time by the sun. Thus it is no surprise that all Islamic *zīj*es deal with this problem.

One way in which the *zīj*es treat the problem is as follows: To calculate the rising time of an arc that stretches from λ to λ' it is sufficient to be able to calculate the rising times of the arcs from 0 to λ and from 0 to λ', for the rising time of the given arc will be the difference of these two rising times. Thus we can solve the problems of rising times if we can calculate the rising time of any given initial arc from 0 to λ.

This is not an easy problem because the rotation of the heavens is around the poles of the equator, not the poles of the ecliptic, and as a consequence equal arcs of the ecliptic do not rise in equal times. However, for the same reason, equal arcs of the equator do rise in equal times and therefore, since 360° rise in a day, 1° of the equator rises every 4 minutes. Thus, if we want to find the rising time of an arc of the ecliptic from 0° to λ° we need only compute how many degrees of the equator have risen in the same time and multiply that number by 4 to obtain the rising time in minutes.

For a locality on the earth's equator the celestial equator will pass through the zenith, so it will be at right angles to the horizon. In this case, the arc of the equator that has risen with the arc of the ecliptic from θ° to λ° is called the *right ascension* of that arc and is written $\alpha(\lambda)$. When the locality is not on the equator then the equator makes an acute angle with the horizon equal to $90° - \phi$, where ϕ is the local latitude. In this case the arc of the equator that has risen is written $\alpha_\phi(\lambda)$ and is called, for obvious reasons, the *oblique ascension* of λ.

Methods for computing $\alpha_\phi(\lambda)$ are found in astronomical treatises of the Babylonians, Greeks and the Islamic world, and one of the fascinating aspects of the history of various schemes for computing them is the survival of very old methods for over 1000 years, long after more sophisticated approaches were available. We shall say no more of some of the oldest methods, which employed arithmetical sequences, but later in this chapter we shall see some of the results of quite sophisticated calculations used by astronomers of the Islamic world. Of course, we can get results that yield a good insight into the problem by means of the spherical models we described earlier.

§4. Stereographic Projection and the Astrolabe

However, of such spherical models al-Bīrūnī wrote that to be of any use they must be of considerable size, but that this very feature of size also means that they are rarely found and difficult to transport and manipulate. As al-Bīrūnī puts it, in his laconic style, "Thus, the difficulty in it corresponds to the good in it". And, whether it was for reasons of convenience or not, it seems that the astronomer Hipparchos of Rhodes, the same man who composed the first table of chords that we know of, wrote a treatise on a method that allows one to represent the surface of a sphere on a plane so that circles on the sphere are represented by circles on the plane. This method is now called stereographic projection (*stereo* = solid, *graphein* = to describe), and although Hipparchos' treatise on the subject is lost, Ptolemy's *The Planisphérium*, written almost 300 years later, has survived.

Stereographic projection, familiar to any student of complex variables, may be described as follows (Fig. 6.4): A great circle on the surface of a sphere, such as the equator, is chosen and the plane π containing it will be the one onto which we will map the surface of the sphere. To effect this mapping, we pick one pole of the great circle, say the south pole, S, and then define for any point $X \neq P$ on the sphere the image point X' on the plane π as the point where the line PX cuts π. Since for any point $X \neq P$ the line PX cuts π in only one point, the image X' is uniquely defined for each $X \neq P$. The effect of this mapping on various points $X, X_1, \ldots$ on the sphere is illustrated in Fig. 6.4.

The utility of the projection lies in the fact that it maps circles on the sphere to straight lines or circles on π and that it preserves angles. Figure 6.4 makes it plain that meridians are mapped onto straight lines through the center of the sphere, and also that the equator, which is on π, is mapped to itself, as well as the fact that points south of the equator (such as P_1) are mapped outside the equator, while points north of the equator are mapped inside it.

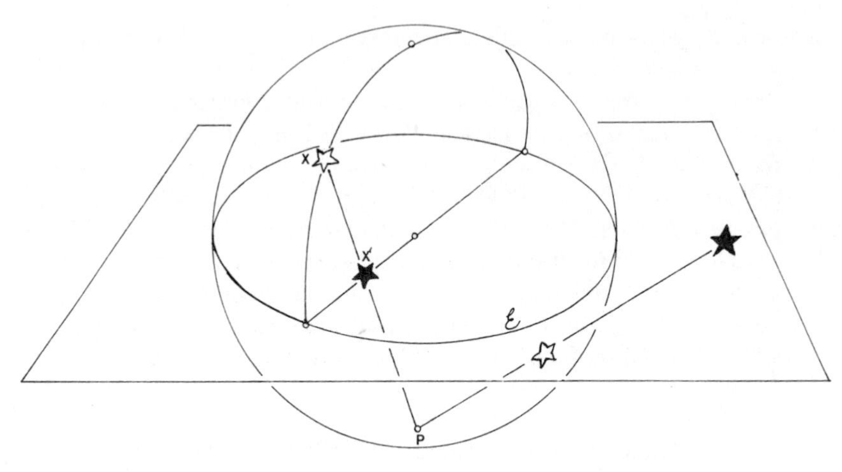

Fig. 6.4
Drawing reproduced courtesy of Paul Mac Allister & Associates.

We do not know whether Hipparchos thought of making any instrument based on stereographic projection, nor does Ptolemy's treatise describe any such instrument. However, the first known treatise on such an instrument was written by Theon of Alexandria late in the fourth century A.D., so at some time prior to this someone had applied stereographic projection not only to the sphere with its stars, but to the system of its important great circles. The result was two disks, one representing the starry sphere and the other the various great circles, in particular the equator and horizon. When the disk carrying the coordinate net of great circles rotates on top of that on which the stars are engraved, one has an anaphoric clock. Such an instrument was housed in the Tower of the Winds, which was built in the Athens marketplace in the mid-first century B.C. and is still standing. However, if a disk carrying pointers to mark the star positions and mounted on a brass framework rotates over a solid disk carrying the coordinate systems then we have the astrolabe. (See Plates 6.2 and 6.3.)

The first reference to an astrolabe that clearly refers to what we understand by that word is in a letter of Synesios, Bishop of Ptolemais, to his teacher Hypatia, the first woman mathematician whose name is known to us. Her father, Theon of Alexandria, did an important edition of Euclid's *Elements* and also wrote on the planispheric astrolabe. In fact, he seems to have equipped Synesios's astrolabe with a sighting device that allowed the user to take the altitude of a star or the sun over the horizon—the basic piece of data fed into the instrument. It was Theon's treatise that was rendered into Syriac by Bishop Severus Sebokht, who made the first known reference to Hindu numerals outside of India, and through the Syriac it became known to the Arabic authors.

Since the stereographic projection maps the sphere (less the south pole) onto the whole plane it is necessary to limit the size of the image, and this is

Plate 6.2. A seventeenth-century Maghribī (western) astrolabe of the standard planispheric variety. The plate of coordinates for the locality is clearly visible below the top star map. Reproduced courtesy of the Trustees of the Science Museum, London.

Plate 6.3. A universal *saphea* on the back of the astrolabe illustrated in Plate 6.2. The grid is based on a projection for the latitude of the equator and is hence independent of the terrestrial latitude. With the usual alidade as shown, the grid is useless. As devised in eleventh-century Muslim Spain the grid must be used together with an alidade fitted with a movable perpendicular cursor. Then one can convert from one orthogonal coordinate system to another. Reproduced courtesy of the Trustees of the Science Museum, London.

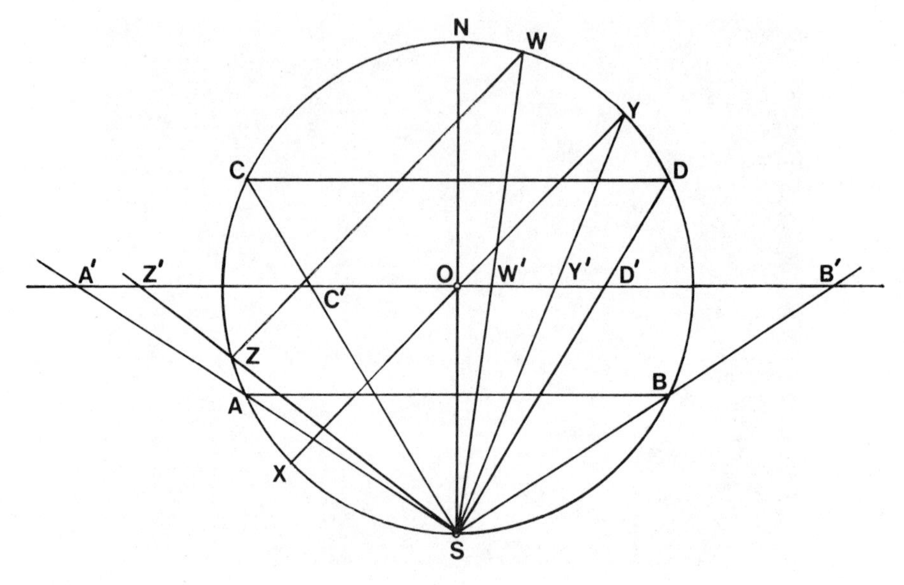

Fig. 6.5

done by mapping only the part of the sphere (and its great circles) above the Tropic of Capricorn, which has a latitude of approximately $-23\frac{1}{2}°$. Thus the whole ecliptic is represented on the plane, along with the equator, the Tropic of Cancer, the horizon (or that part of it above the Tropic of Capricorn) and the circles of equal altitude above the horizon and parallel to it. (These latter are still known by their Arabic name, *almucantars*.) Figure 6.5 shows the resulting map of the coordinate circles in section, where capital letters represent points on the sphere and the letters with dashes are their images under stereographic projection. Solid lines indicate the equator and circles parallel to it, so AB is the section of the Tropic of Capricorn, while CD is the section of the Tropic of Cancer. XY is the section of the local horizon and the line ZW, parallel to it, is the section of an almucantar. Notice that since, in the case shown, the southern edge of the horizon is below the Tropic of Capricorn, it is not mapped onto the plane, so that, for example, X has no image.

Figure 6.6 shows the images of this mapping for the latitude 30°. The top of the figure represents the south. The outer circle represents the Tropic of Capricorn; the next smaller circle, concentric with it, represents the equator, and the inner one of the three concentric circles is the Tropic of Cancer. The center of these circles is the north celestial pole and it is represented by a small circle.

The circle south of the pole represents the zenith for latitude 30°, and the circles clustering around it are almucantars at intervals of 20°, extending down to the horizon itself. (Anything outside the horizon circle is

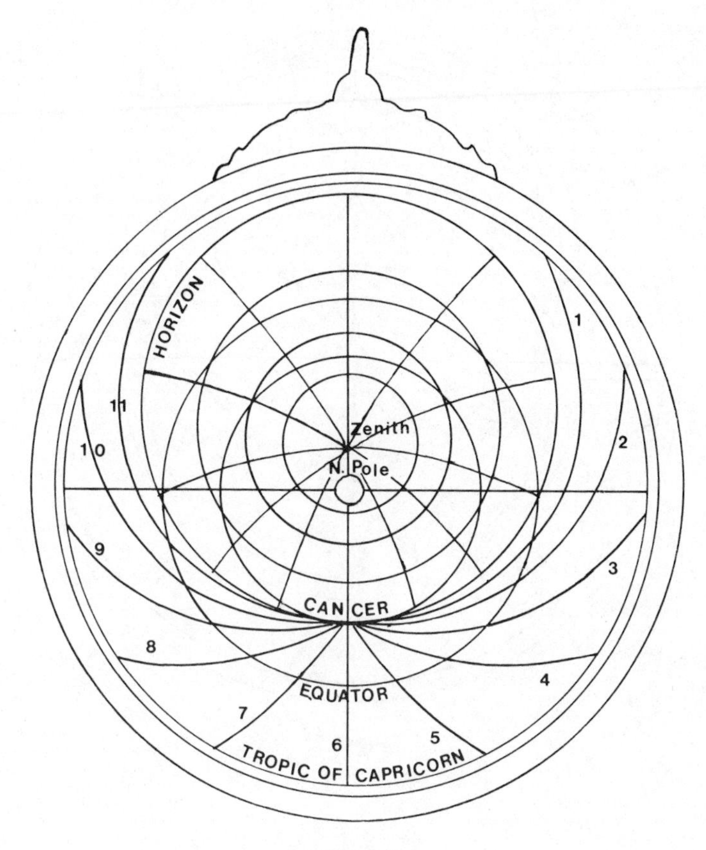

Fig. 6.6

below it on the globe and thus invisible.) As we said, the southern extremity of the horizon is below the Tropic of Capricorn, and therefore the horizon (as well as some of the lower circles parallel to it) is not represented by a complete circle. The curves (actually circular arcs) joining the zenith to the horizon are the images of the altitude circles, marking out intervals of 10° on the horizon, proceeding clockwise from the north point N, which is 0°.

On top of the plate just described is a circular star map, of the same diameter as the lower plate, which shows certain important stars above the Tropic of Capricorn as well as the circle of the zodiac. Only pointers indicating the positions of the stars and a supporting fretwork are on the star map, and the rest of the brass plate is cut away to allow the user to see the circles on the lower plate. A post goes through the center both of the star map and the lower plate, and the rotating of the star map around the post imitates the rotation of the heavens around the north pole. Figure 6.7 shows the star map, called *rete* (= "net") by the Latin astrolabists, a less colorful term than the Greek and Arabic equivalents, which mean "spider".

Fig. 6.7

The astrolabe is used as follows. It is hung from a cord held in the hand, so it is hanging vertically. The ruler on the back pivots around the central post and we may sight along it to a particular star and read the altitude of the star from the scale of degrees around the rim. Suppose we find Spica is 16° above the horizon in the southwest. Rotate the star map until the pointer for Spica is in the southwest, 16° above the horizon. (It will then be on the eighth almucantar of the coordinate plate.) Now the astrolabe shows all stars in their correct positions, and gives the altitude and azimuth of any star on the plate. It shows, in particular, which stars are just rising/setting (i.e. have their pointers on the eastern/western horizon) and which are below the horizon and therefore invisible.

The earliest surviving Arabic treatise on the use of the astrolabe is one written by ʿAlī b. ʿIsā, a scientist who flourished around 830 and participated in al-Maʾmūn's survey to determine the circumference of the earth. In addition, he took part in astronomical observations both in Baghdad and in Damascus, so he must have had the experience necessary to write a treatise dealing with a great variety of uses of the astrolabe—among them the following:

1. Determination of the longitude of the sun in the ecliptic.
2. Azimuth and altitude of any Star.
3. Determination of ascendant, descendant, houses and other astrological uses (for casting horoscopes).
4. Length of daylight or night, length of the unequal hours.
5. Time of day in equal or unequal hours.

We shall concern ourselves with the last of these.

§5. Telling Time by Sun and Stars

To understand ʿAlī's directions for telling the time of the night or day one must know the two systems for recording time in the ancient or medieval world. The first system, the popular one, divided each day and night into twelve equal parts, each of which was one seasonal (or "temporal") hour. Clearly the hours of the day would be longer in summer and shorter in winter. Only on the equator, where all days and nights are equal, would the hours not vary with the seasons. Elsewhere, it is only on the dates of the spring and fall equinoxes, when the sun is on the equator, that the daytime hours are equal to the nighttime hours. Thus, in the second system, where every day–night has 24 hours of equal length, the hours are referred to as *equinoctial hours*. We shall say no more about these, although one can also determine them readily with the astrolabe.

As for the problem of determining the seasonal hours Fig. 6.8 shows the celestial sphere with a horizon $\widehat{ABG}$. On a given day of the year, the sun will

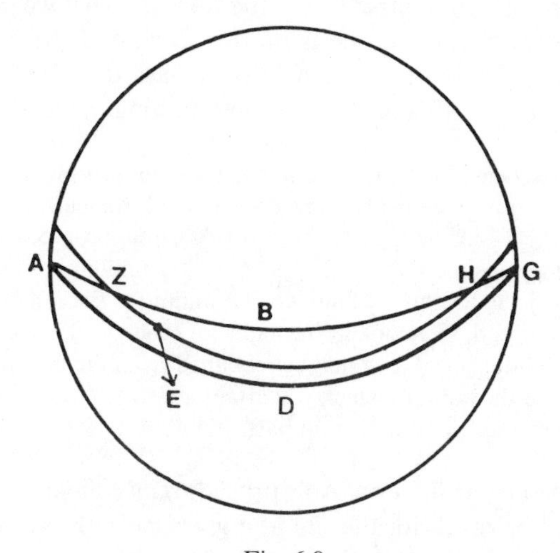

Fig. 6.8

describe a circle parallel to the equator $\widehat{ADG}$. Let the part of the sun's day-circle below the horizon be $\widehat{ZEH}$, so that this is the path of the sun during the night. If $\widehat{ZE} = \frac{1}{12}\ \widehat{ZEH}$ then the arc the sun travels in $\frac{1}{12}$ the night is $\widehat{ZE}$. The corresponding point E on the other arcs, one for each night of the year, forms a continuous curve on the sphere, and some of the Muslim astronomers realized that the curve is not a circle. Hence, its image will not be a circle on the astrolabe. However, if we take the night arcs only when the sun is on the Tropic of Capricorn, the Tropic of Cancer or the equator, then we get only three points for each hour. The images of these three points on the astrolabe determine the arc of the unique circle joining them, and it is this arc that is labelled 1, with the successive arcs labelled 2, ..., 11, and the eastern horizon labelled 12, for when the sun is there it is the end of the twelfth hour of the night, i.e. it is dawn. Of course, these circular arcs are only approximations to the real curves, but most applications of mathematics demand some approximation, and the circular arc is a reasonable one. This is how the lines for the night hours are drawn on the astrolabe.

Now, if we wish to tell what hour of the night it is, we can proceed as follows. Either from tables or from a scale on the back of the astrolabe, we find out where the sun is in the zodiac on the day in question and mark that point on the zodiac circle appearing on the star map. This sets the sun in the right position relative to the stars for that day. Now find in the sky any of the 30 or so stars of those on the star map, and take its altitude, noting whether it is to the east or west of the meridian. Then set the star map of the astrolabe so that the pointer corresponding to the star observed is on its circle of altitude and east or west of the meridian as the case may be. Now the star pointers are in the correct position relative to the horizon line, and so the sun, correctly placed among the stars, is in a place faithfully representing its position in the sky (of course below the horizon). All we need do now is look to see what hour-line the sun is on (or near), and the number on that line tells us how many hours of the night have passed.

We may now quote ʿAlī's method for determining the time of day:

> The nadir is the degree (of the ecliptic) that lies exactly opposite the sun's position, the one seven signs after the one in which the sun is located. (ʿAlī begins counting "1" in the sign of the sun.) And one proceeds continuously until he finally attains the seventh sign from his beginning point, which is then the position of the nadir. Put the point corresponding to the sun on the altitude which you have found, then look at the nadir (of this point), which falls on (or near) one of the hour lines. It should be reckoned from the beginning point of the counting, and the point to which you attain, on which the nadir falls, is the amount of hours and fractions thereof past.

In order to shed some light on ʿAlī's procedure, we have drawn in Fig. 6.9 the equator and horizon, with the sun at a given point P on the ecliptic. We consider a straight line through the center of the sphere and P. Since P and the center are on the plane of the ecliptic, the point P*, where the line joining

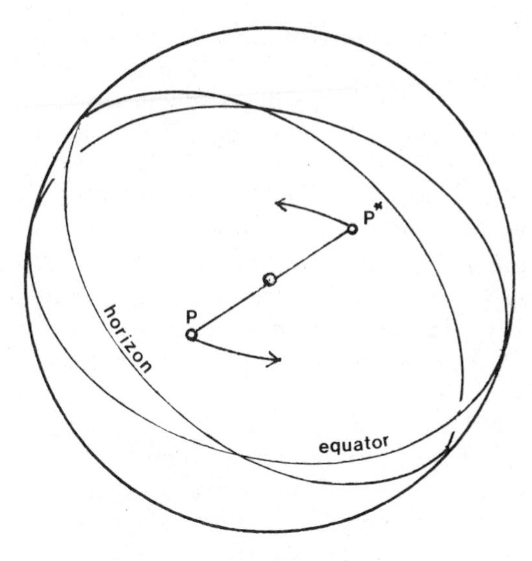

Fig. 6.9

P and the center cuts the sphere, is on the ecliptic and is the nadir. Now imagine P rotating in the direction shown, parallel to the equator. As it does, P* rotates an equal amount in the same direction. Moreover, because a great circle contains the diameter through any of its points, P will be on a given great circle exactly when P* is. Our conclusion is that P will sink below the horizon precisely when P* rises above the horizon, and that whatever fraction of its daily circle, parallel to the equator, the sun, P, traces out to reach the horizon, its nadir, P*, traces out precisely the same fraction of its parallel circle in reaching the horizon. Thus if P* falls on hour-line n, indicating $12 - n$ hours remain of the night, then, for P, $12 - n$ hours remain of the day.

Since the astrolabe is an analogue computer, which faithfully models circular arcs and angles in the heavens, it can be used to solve any problem of spherical astronomy. However, its accuracy is necessarily limited by the skill with which it was fabricated, and it does not provide the most elegant solutions for all problems.

In fact no single method does (which is part of the charm of spherical astronomy), but spherical trigonometry, with its powerful and often easily-stated rules, was one source of several lovely solutions, and it is to the development of this subject in the Islamic world that we now turn.

§6. Spherical Trigonometry in Islam

There are three astronomers whose careers span the period during which most of spherical trigonometry was developed and who, themselves, were responsible for the most important results. The first of these is Ḥabash al-

Ḥāsib, who was a contemporary of the great Arab scientist al-Kindī, and who was one of the astronomers active in Baghdad under the patronage of the caliph al-Ma'mūn. The second is the astronomer Abu l-Wafā' al-Būzjānī, one of the ornaments of the Būyid court in the middle and late tenth century, whose life and contributions to geometry and trigonometry we have already recounted. The third is the prince Abū Naṣr Manṣūr ibn ʿIrāq who was both teacher and patron of al-Bīrūnī in the latter part of the tenth century. In his *Keys to the Science of Astronomy* al-Bīrūnī gives a lively account of the controversies, misunderstandings and accusations that accompanied the disputes over priority in the discovery of some important theorems in spherical trigonometry—particularly as they concerned the latter two astronomers mentioned above.

In the course of his account, al-Bīrūnī has some unkind things to say of Abu l-Wafā', but, despite al-Bīrūnī's low estimate of Abu l-Wafā''s character, we shall follow the proofs of two major theorems as Abu l-Wafā' presented them in his astronomical handbook *Zīj al-Majisṭī.*

The first of these results is the following "Rule of Four Quantities" (as it came to be known later in the Latin West):

> If ABG and ADE are two spherical triangles with right angles at B, D respectively and a common acute angle at A then $\mathrm{Sin}(\widehat{BG}):\mathrm{Sin}(\widehat{GA}) = \mathrm{Sin}(\widehat{DE}):\mathrm{Sin}(\widehat{EA})$ (Fig. 6.10).

Abu l-Wafā''s proof is as follows: Since the arcs $\widehat{AB}$ and $\widehat{AG}$ are great circles the planes containing these arcs both contain the center of the sphere and so intersect in a diameter d of the sphere. From G and E drop perpendicular

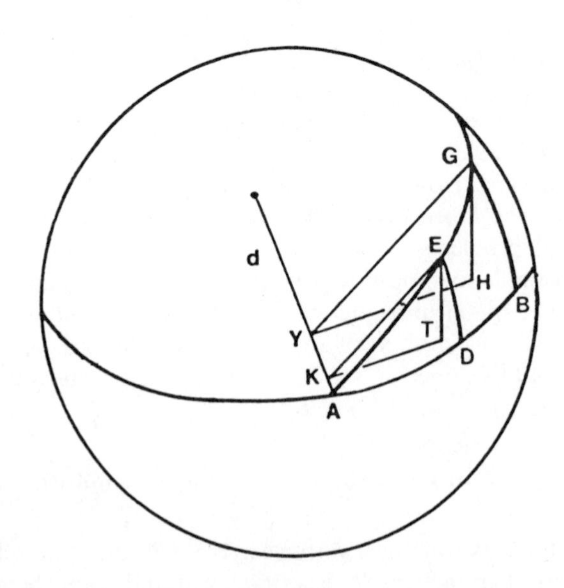

Fig. 6.10

straight lines inside the sphere, GH and ET onto the plane containing $\widehat{AB}$, and in the plane containing $\widehat{AG}$ draw perpendicular straight lines GY and EK onto the diameter *d*. Then it is an easy consequence of Euclid's *Elements*, XI,11 that YH and KT are also perpendicular to d. Thus the angles GYH and EKT are equal, and so Δ(GHY) is similar to Δ(ETK). Hence TE : EK = HG : GY. But, EK = Sin($\widehat{EA}$) and GY = Sin($\widehat{GA}$), while TE = Sin($\widehat{ED}$) and HG = Sin($\widehat{BG}$), and substituting these into the previous proportion yields the conclusion of the theorem.

An important application of the Rule of Four Quantities is found in Abu l-Wafā''s derivation of the Law of Sines for spherical triangles. Its discovery simplified many problems concerned with arcs on the sphere and marked the full emergence of spherical trigonometry because it was the first theorem to use spherical angles. Other theorems used spherical triangles, but they dealt with only the sides.

Given the importance of the theorem it is not surprising that several authors claimed credit for its discovery. Simultaneous discovery is not uncommon in mathematics, since most problems and current methods are known to all workers, and the case of the spherical law of sines seems to be another instance of it. However, it appears that Abu l-Wafā' was the first to publish it, in his *Zīj al-Majisṭī*, and to use it, so perhaps the greater part of the credit for this important advance goes to him.

The Law of Sines for spherical triangles says: If ABG is a spherical triangle with sides *a*, *b*, *g* opposite the angles A, B, G then

$$\frac{\mathrm{Sin}(a)}{\mathrm{Sin}(\mathrm{A})} = \frac{\mathrm{Sin}(b)}{\mathrm{Sin}(\mathrm{B})} = \frac{\mathrm{Sin}(g)}{\mathrm{Sin}(\mathrm{G})}.$$

Abu l-Wafā''s proof is as follows: In Fig. 6.11 let the spherical triangle,

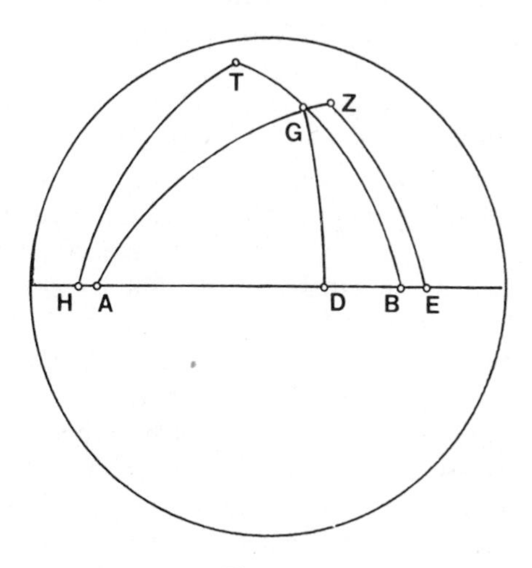

Fig. 6.11

Δ(ABG), be given, and let GD be an arc of a great circle perpendicular to $\widehat{\text{AB}}$. Extend $\widehat{\text{AB}}$ and $\widehat{\text{AG}}$ to $\widehat{\text{AE}}$ and $\widehat{\text{AZ}}$, both of them quadrants, and extend $\widehat{\text{BA}}$ to $\widehat{\text{BH}}$ and $\widehat{\text{BG}}$ to $\widehat{\text{BT}}$, both quadrants. Then A is a pole for the great circle $\widehat{\text{EZ}}$ and B a pole for the great circle $\widehat{\text{TH}}$. Thus, by the second part of Basic Fact 2 in Section 1 the angles E, H are both right, and the right-hand side triangles ADG and AEZ are spherical right triangles with a common angle at B. Thus by the rule of four quantities

$$\frac{\text{Sin}(\widehat{\text{DG}})}{\text{Sin}(b)} = \frac{\text{Sin}(\widehat{\text{ZE}})}{\text{Sin}(\text{ZA})} \quad \text{and} \quad \frac{\text{Sin}(\widehat{\text{DG}})}{\text{Sin}(a)} = \frac{\text{Sin}(\widehat{\text{TH}})}{\text{Sin}(\widehat{\text{TB}})}.$$

However, as remarked earlier, A and B are poles of $\widehat{\text{ZE}}$ and $\widehat{\text{TH}}$ respectively, so, by definition of spherical angles, $\widehat{\text{ZE}} = \sphericalangle \text{A}$ and $\widehat{\text{TH}} = \sphericalangle \text{B}$, and so we may rewrite the above equalities as

$$\frac{\text{Sin}(\widehat{\text{DG}})}{\text{Sin}(b)} = \frac{\text{Sin(A)}}{R} \quad \text{and} \quad \frac{\text{Sin}(\widehat{\text{DG}})}{\text{Sin}(a)} = \frac{\text{Sin(B)}}{R}.$$

Eliminating $\text{Sin}(\widehat{\text{DG}})$ from these two equations, we obtain $\text{Sin}(a)/\text{Sin(A)} = \text{Sin}(b)/\text{Sin(B)}$. The proof of the remaining equality is entirely similar and the theorem is thereby proved.

§7. Tables for Spherical Astronomy

In the chapter on trigonometry we mentioned that the auxiliary tables of trigonometric functions were computed by a series of astronomers from the ninth to the fourteenth centuries, and we pointed out that one of the principal uses of such tables was to aid in the computation of tables of functions important for spherical astronomy. Our aim in this section is to provide details on these latter tables, for they are one of the crowning achievements of numerical methods in medieval mathematics.

A good example is the history of tables of oblique ascensions, which were defined earlier in this chapter, and we shall begin with the tables of Ibn Yūnus. In his *Ḥākimī zīj* this Egyptian astronomer tabulated these ascensions for each degree of the ecliptic, and for each degree of latitude from 1° to 48°, to minutes of arc. This means the computation of nearly 18,000 entries. D. A. King, who has investigated these tables, reports that for the latitude of Cairo (30°) only one of the first 90 entries was in error by 1 minute. However, for the table as a whole, King found that about one-third of the entries were incorrect by 1 minute, and for a few special cases, such as latitude 40°, the proportion of entries in error by 1 minute increased to two-thirds. There are also isolated errors of 2 or 3 minutes.

Two facts emerge from the information above: Ibn Yūnus performed an impressive piece of numerical mathematics and, as the fair number of small but significant errors makes clear, he used some mathematical methods to interpolate between accurately computed values. However, as King's analysis of the errors makes plain, the interpolation procedure was not linear, so the tables of Ibn Yūnus are evidence that in the Islamic world during the late tenth century there were large-scale computational endeavors, using non-trivial mathematical formulas and methods. (Apropos of the question why Ibn Yūnus stopped at latitude 48° King conjectures that perhaps his feelings were the same as those expressed by Abū Naṣr expressed in his auxiliary tables, namely, "I have made it (the table) for latitudes 1° to 45° since amongst the inhabitants of the places whose latitude is greater than this there is scarcely anybody who studies this sort of thing or even thinks about it".)

Evidently, as the centuries passed, people of the northern latitudes became more interested in such matters since Naṣīr al-Dīn al-Ṭūsī, in his *Ilkhanī zīj*, calculates the oblique ascensions for each degree of the ecliptic, but this time for all latitudes from 1° to 53°, and not to minutes but to seconds. Then, some century and a half later al-Kāshī calculated the oblique ascensions, again to seconds, up to latitude 75°, and his patron, Ulūgh Beg, calculated these ascensions to latitude 50°, but to thirds. The work of al-Kāshī and Ulūgh Beg exemplifies the quality of the best Muslim achievements in the production of accurate, extensive scientific tables of functions arising in spherical astronomy.

Our second example of tables of functions from spherical astronomy are the tables for timekeeping. Ibn Yūnus compiled the first extensive collection of such tables for the latitude of Cairo, for the purpose of determining the time by means of the sun and stars for civil or astronomical uses, as well as regulating the times of the five daily Muslim prayers. These times were defined in terms of the position of the sun relative to the horizon, hence the composition of such tables is an exercise in spherics applied to astronomy, i.e. in spherical astronomy. This science of timekeeping (*ʿilm al-miqāt* in Arabic) gave rise to a group of astronomers who were associated with major mosques and whose duty it was to tell the *muezzin* when to call the faithful to prayer.

Collections of such tables were quite large, that of Cairo containing 200 pages of 180 entries each. The following is a survey of the contents of this corpus as it is found in the work attributed to Ibn Yūnus with the descriptive title *Very useful tables for finding the time since sunrise, the hour-angle and the azimuth of the sun from the altitude of the sun*. These tables fall into these main classes:

1. Auxiliary Tables of Functions for Spherical Astronomy

Among the 13 tables in this group are tables giving, for each degree of longitude λ of the sun, the declination of the sun, $\delta(\lambda)$, the length of daylight for the day when the sun has longitude λ, and the height of the sun when it is due south, due east or due west.

جدول فضل الدائر لنصف النهار من ارتفاع سمت القبلة ٥

العدد	حمل / سنبلة		ثور / اسد		جوزا / سرطان		ميزان / حوت		عقرب / دلو		قوس / جدي	
[illegible]	[illegible]	[illegible]	[illegible]	[illegible]	[illegible]	[illegible]	[illegible]	[illegible]	[illegible]	[illegible]	[illegible]	[illegible]

جدول يعرف منه حصة الشفق من ارتفاع [illegible] ٥

العدد	حمل / سنبلة		ثور / اسد		جوزا / سرطان		ميزان / حوت		عقرب / دلو		قوس / جدي	
[illegible]	[illegible]	[illegible]	[illegible]	[illegible]	[illegible]	[illegible]	[illegible]	[illegible]	[illegible]	[illegible]	[illegible]	[illegible]

Plate 6.4. Two tables from the corpus of tables for timekeeping that was used in Cairo throughout the medieval period. These display the time before midday when the sun is in the direction of Mecca, and the duration of evening twilight. Values are given in equatorial degrees and minutes for each degree of solar longitude. Taken from MS Dublin Chester Beatty 3673, fols. 8^{v}–7^{r}. Reproduced courtesy of the Trustees of

2. Tables of Time Since Sunrise and the Hour Angle

These times are tabulated as functions of the longitude (on the ecliptic) of the sun on the day in question and the altitude of the sun at the instant it is observed. (One way to determine the altitude of the sun is by the sighting device on the back of an astrolabe.) (See Plates 6.4 and 6.5 for an example of such tables.) In his study of these tables D. A. King points out that the often-repeated assertion that Ibn Yūnus was the first to propound the so-called prosthaphairesis formula

$$\cos(\theta)\cdot\cos(\delta) = \tfrac{1}{2}\cdot[\cos(\theta+\delta)+\cos(\theta-\delta)]$$

to facilitate computing the hour-angle from the altitude of the sun is a misunderstanding by the nineteenth century French historian J.-B. Delambre.

3. Tables of the Azimuth of the Sun

These tabulate the azimuth as a function of the same arguments as the previous tables used, namely solar longitude and altitude. King remarks that these values seldom deviate from the true values by more than 1 in the second digit, and they are calculated for each degree of solar altitude up to a maximum of 83°. (This is approximately the maximum for Cairo.)

4. Tables of the Altitude of the Sun for Certain Azimuths

The other argument, beside azimuth, is longitude of the sun, and the composition of these tables shows a certain excessive zeal for computing, since it is hard to think of any purpose for which they would be really convenient, let alone necessary. Other tables, very much in the same spirit as the foregoing, are the

5. Tables for Orienting Ventilators

Ventilators were used on the top of houses to draw cooling winds down into the buildings, and certainly any book that could tell the residents of a city as hot as Cairo how to do this could lay justifiable claim to the description "Very Useful". Indeed, the Iraqi traveller ʿAbd al-Laṭīf al-Baghdādī, who visited Egypt around the year 1200, wrote of the Egyptians that they

> ... make the openings of their houses exposed to the agreeable winds from the north. One sees hardly any houses without ventilators. These ventilators are tall and wide, and open to every action of the wind; they are erected carefully and with much skill. One can pay between one hundred and five hundred dinars for a single ventilator, but small ones for ordinary houses cost no more than one dinar each.

However, Ibn Yūnus' *Very Useful Tables* did not tell the Cairenes in which direction they should orient their ventilators. Rather, the book gave the altitude of the sun when it was in the azimuth of the direction everyone used for orienting ventilators. This is strange enough, but even

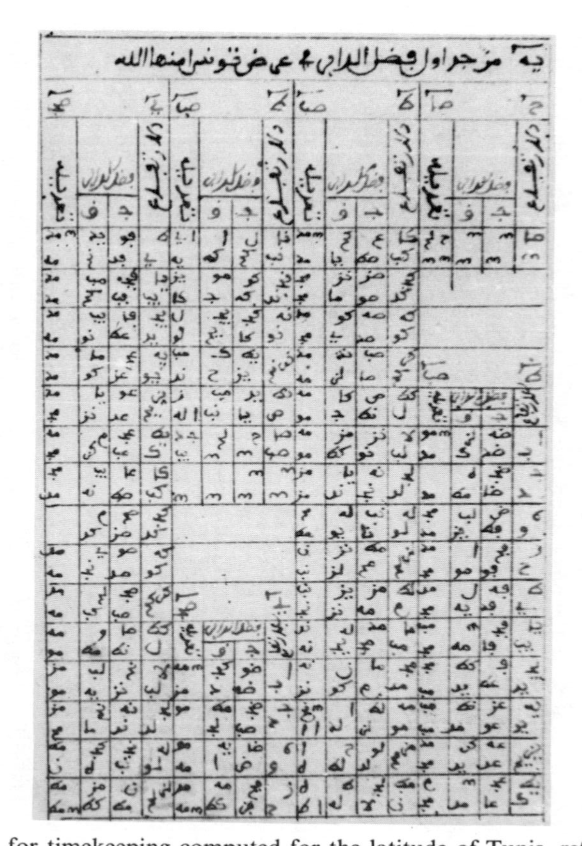

Plate 6.5. Extract from some anonymous fourteenth (?) century tables for timekeeping computed for the latitude of Tunis, referred to as "Tunis, the protected (by Allah)" in the heading of the table on the left. The tables display the time until noon as a function of solar meridian altitude and instantaneous altitude. (Taken from MS Berlin Staatsbibliothek, (Ahlwardt 5754 fols. 23^v–24^r (We 1138).)

stranger is the direction the Cairenes used (and Ibn Yūnus prescribed) for orienting ventilators, namely the direction of the rising sun at the winter solstice, which Ibn Yūnus calculated to be 27°30′ south of east, whereas modern data suggest that the optimal alignment for the winds in Cairo is about 70° south of east. However, recent studies by D. A. King show that in the medieval Islamic world a tradition of folk astronomy associated the direction of the rising sun at the winter solstice was associated with certain winds. Thus this part of Ibn Yūnus' treatise shows a blend of folk astronomy with sophisticated calculations that is one of the pleasures of this field of historical investigation.

6. **Tables of the Duration of Morning and Evening Twilight**
   Twilight was defined in terms of the depression of the sun below the horizon, and its determination is important because it defines times appropriate for morning and evening prayers. Also, the use of these tables, in connection with tables of the length of daylight, allowed the computation of the length of real darkness for a given day, and Ibn Yūnus tabulates this function.

7. **Tables for Afternoon Prayer**
   Although conventions on the time of afternoon prayer in the Islamic world varied, Ibn Yūnus uses the convention that afternoon prayer begins at the time after noon when the shadow of an upright rod in the ground equals the length of its noon shadow plus the length of the rod. Ibn Yūnus tabulates, for each degree of longitude of the sun, the altitude of the sun at the beginning of afternoon prayer. (The permitted time for this prayer ends just before sunset.)

8. **Tables of Corrections for Horizontal Refraction**
   In his *Optics* Ptolemy considered in a qualitative way the effects of atmospheric refraction, especially at the horizon, but a table found in one of the manuscripts of Ibn Yūnus' *Very Useful Tables* makes it clear that Muslim astronomers of the medieval period tried to get quantitative estimates of the effect of refraction. The table under discussion applies to sunrise and sunset, but it appears in only one of the manuscripts and contains blunders inconceivable in one as expert in spherical astronomy as Ibn Yūnus. Thus, King feels the table was not composed by Ibn Yūnus but was added, by a much less competent writer, on the basis of remarks by Ibn Yūnus which he only half-understood.

The above, then, are descriptions of some of the tables found in the *Very Useful Tables*, a corpus of tables that served Egyptian astronomers and timekeepers until the nineteenth century.

In later centuries even more ambitious tables were undertaken. For example, in 1250 the Egyptian astronomer Najm al-Dīn al-Miṣrī tabulated the time since the rising of a Star as a function of three quantities: (1) the maximum altitude of the Star; (2) the instantaneous altitude of the Star; and

(3) half of the arc of visibility. These tables are computed for all declinations of the Star and for all terrestrial latitudes and contain over a quarter of a million entries.

In the next century, in Damascus, the time-keeper at the Umayyad mosque, Muḥammad al-Khalīlī, tabulated practically all the functions Ibn Yūnus tabulated, but for the latitude of Damascus and for a different value of the obliquity of the ecliptic. Perhaps it was the labor of doing all over again what Ibn Yūnus had done for Cairo that inspired al-Khalīlī to compose the auxiliary tables we mentioned in the chapter on trigonometry, tables which one could use to solve the standard problems of spherical astronomy and which, therefore, would allow the user to draw up a similar set of time-keeping tables for his latitude. In the following section we shall see more of al-Khalīlī's universal solutions to mathematical problems arising within the context of Islam.

§8. The Islamic Dimension: The Direction of Prayer

The problem of finding the direction of Mecca relative to a given locality was a product of the religion of Islam, for Mecca is the site of the Kaʿba, the most sacred spot in the Islamic world, and it is the direction to which Muslims must turn to say their five daily prayers. This direction is called, in Arabic, *al-qibla*, and the problem of its determination is an important one for Muslims. Accordingly, many of Islam's greatest scientists devoted some attention to its solution.

One of the greatest of these, al-Bīrūnī, wrote near the end of his definitive work on mathematical geography, *The Determination of the Coordinates of Cities*, as follows:

> Though the determination of position is an end in itself, which satisfies an investigator, it is our duty to find an application for such a determination which is beneficial to the populace of the whole region whose longitude and latitude we have surveyed, or to a particular section of it exclusively. Let the universal benefit be the determination of the azimuth of the *qibla*.

It is typical of al-Bīrūnī's tolerance for religions other than his own that he also mentions the duty of the Jews to face Jerusalem and of the Christians to face east, and he says his techniques will also be useful to them, and "I have no doubt that is useful also to people of all faiths".

In any case, al-Bīrūnī presents four methods for solving this problem. Although a detailed exposition of one of these is beyond our purpose here, we shall describe the problem and illustrate how it may be solved by spherical trigonometry.

Consider the problem on the earth's surface. Figure 6.12 illustrates the situation for a locality northwest of Mecca, where P is the north pole, Z the

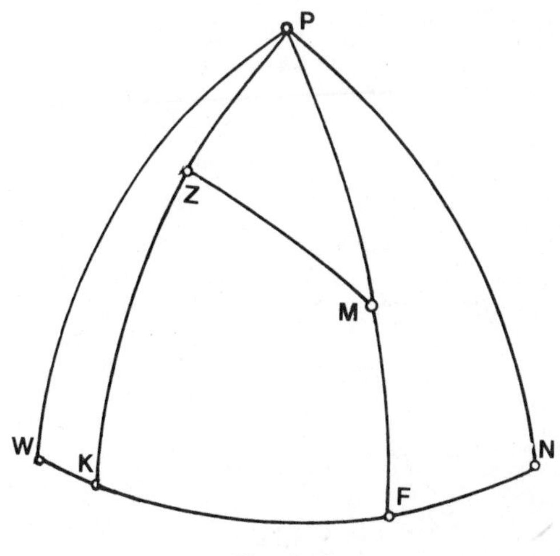

Fig. 6.12

locality in question, M the position of Mecca, and WKFN the equator. (Since we shall consider only arcs in what follows, and never straight lines, we will write XY for $\widehat{XY}$ without any ambiguity.) Thus PZ and PM are the local meridian and that of Mecca, respectively, and $\measuredangle$PZM is the local azimuth of Mecca, that is the *qibla*. Moreover, ZM is the great circle distance (in degree) from the locality to Mecca. Since KZP is the local meridian, we will be facing north if we stand at Z and orient ourselves so we face along ZP. Then, if we turn towards our right through angle $\measuredangle$PZM we will be facing Mecca, since ZM is the shortest route to that city. Hence, to calculate the *qibla* of Mecca we must calculate $\measuredangle$PZM.

Clearly, if we are to find this angle we must know where we are and where Mecca is, that is, we must know both our latitude, ϕ, and that of Mecca, ϕ_M, as well as the two longitudes, or at least their difference $\Delta\lambda$ (and whether Mecca is east or west of the local meridian). From the latitudes we can determine their complements, PZ and PM. Thus, if we are to find the *qibla* we must know PZ and PM as well as $\measuredangle$ZPM, that is, two sides and the included angle in $\triangle$(ZPM). However, a spherical triangle is determined by its two sides and the included angle, so these data are sufficient to solve the problem.

It is apparent, however, that we cannot solve the problem by a single application of the Sine Theorem to the spherical triangle PZM, for we do not know both an angle and the side opposite to it. However, there is an approach that applies the Sine Theorem to a series of spherical triangles. This was given by Ibn Yūnus without any justification, but al-Bīrūnī both stated and justified it in his *Mas'udic Canon*. It is this account, as given in King (1979), which we shall follow.

First, however, the reader should recall that for any locality the altitude of

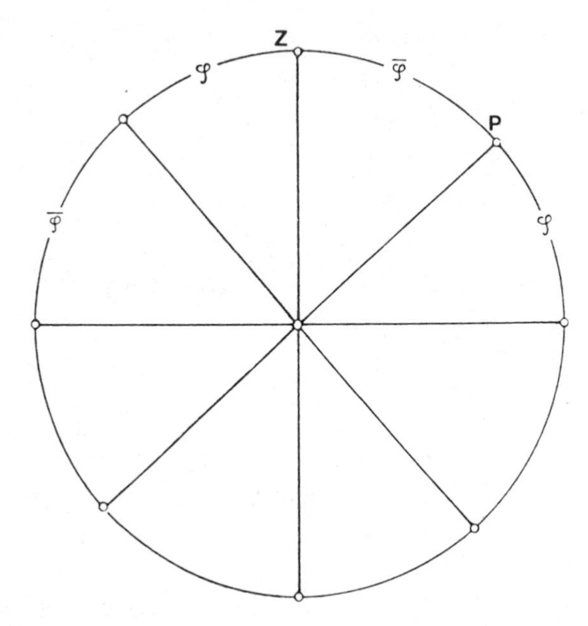

Fig. 6.13

the visible pole (P) above the horizon is equal to ϕ, the local latitude, and the distance, measured in degrees, along the meridian between the zenith (Z) and the visible pole is equal to $\bar{\phi}$. These and other relations are illustrated in the circle of latitudes shown in Fig. 6.13.

If we call the great circle whose pole is a point X "the horizon of X" then an easy consequence of the definition of the poles of a great circle is that X is a point on the horizon of Y if and only if Y is a point on the horizon of X. This principle has two immediate consequences, which we leave as exercises: (P1) The horizons of two nonantipodal points intersect in the poles of the great circle containing them; and (P2) if Y lies on the horizon of X then X, its horizon and its antipode divide the horizon of Y into four quadrants.

We now introduce al-Bīrūnī's diagram for finding the *qibla*, as it appears in the *Masʿudic Canon*, although we have put the locality to the northwest rather than to the northeast of Mecca. In Fig. 6.14 the circle KSN is the horizon of some locality viewed from above, Z represents the local zenith, S the south and N the north, so NZS is the local meridian. The point M is the zenith of Mecca, so that NK or (equivalently) KS is the arc we need to find to know the *qibla*. Now let GFL be the horizon of Mecca, where F is an intersection of the horizon with the local meridian, and let MHJ be the horizon of F. Finally, draw the great circle MPL, where P is the north celestial pole.

Since M is a pole of GLJ, all three of $\angle$MLG, MJ and ML are 90°, and since F is a pole of MHJ both $\angle$FHM and FH are 90°. Also PN = ϕ,

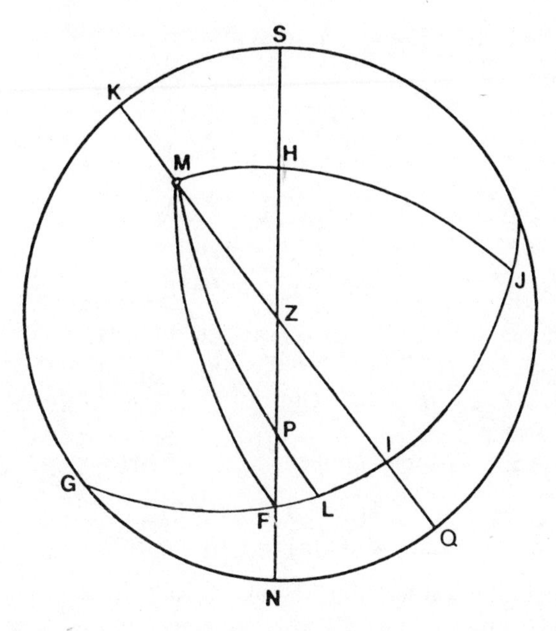

Fig. 6.14

$\measuredangle MPH = \Delta(\lambda)$ and $PL = \phi_M$, the latitude of Mecca, so $MP = 90 - \phi_M$. By the Sine Theorem applied to $\Delta(MPH)$:

$$\frac{\sin(MP)}{\sin(MH)} = \frac{\sin(\measuredangle FHM)}{\sin(\measuredangle MPH)},$$

which, using the above relations and the fact that $\sin\theta = \cos(\bar{\theta})$, becomes,

$$\frac{\cos(\phi_M)}{\cos(HJ)} = \frac{\sin(90°)}{\sin(\Delta\lambda)}.$$

All quantities in this relation except cos(HJ) are known so cos(HJ), and hence HJ, can be obtained. Then $\measuredangle F = HJ$ is known and so is MH, which is equal to $90° - HJ$.

Next, the Sine Theorem applied to $\Delta(PLF)$ yields:

$$\frac{\sin(\measuredangle F)}{\sin(\measuredangle PLE)} = \frac{\sin(PL)}{\sin(PF)}$$

and, substituting the known values into this yields

$$\frac{\sin(\measuredangle F)}{\sin(90°)} = \frac{\sin(\phi_M)}{\sin(PF)}$$

so that sin(PF), and hence PF, is known. Then $FN = \phi - PF$ is known and thus its complement FZ is known.

Next, by the Rule of Four Quantities applied to $\triangle(FZI)$ and $\triangle(FHJ)$, we conclude that:

$$\frac{\sin(FZ)}{\sin(ZI)} = \frac{\sin(FH)}{\sin(HG)}$$

which, on substituting in known values, becomes

$$\frac{\sin(FZ)}{\sin(ZI)} = \frac{\sin(90^\circ)}{\sin(HJ)}$$

and, since all quantities except sin(ZI) are known we may determine ZI and, hence, $IQ = \overline{ZI}$. But, (P1) applied to the horizons of M and Z implies that G is a pole of KMZIQ, the altitude circle of Mecca, and hence that $\measuredangle G = IQ$, so $\measuredangle G$ is known.

Finally, the Sine Theorem applied to $\triangle(GFN)$ yields

$$\frac{\sin(\measuredangle G)}{\sin(\measuredangle F)} = \frac{\sin(FN)}{\sin(GN)}$$

and therefore GN is known. But we have already remarked that $GQ = 90^\circ$, so $NQ = \overline{GQ}$ is known. Since KS = NQ the *qibla* of Z is determined.

Al-Bīrūnī's procedure is just one of the many solutions proposed to the problem of finding the direction of prayer in Islam. Another solution involved composing tables showing the *qibla* for some list of localities. An especially impressive achievement was the set of tables with 2880 entries composed by the fourteenth-century *muwaqqit* (time-keeper for a mosque) Muḥammad al-Khalīlī, which showed the direction of Mecca relative to the local horizon for each locality of latitude $10^\circ, 11^\circ, \ldots, 56^\circ$ and $33\frac{1}{2}^\circ$ (the latitude of Baghdad and east or west of Mecca by $1^\circ, 2^\circ, \ldots, 60^\circ$. Solutions included methods of approximation, descriptive geometry, solid geometry, trigonometry and the construction of sets of tables ranging from a few dozen entries to many thousands. The survival of this multitude of methods over hundreds of years suggests that the history of ancient and medieval spherics is not one of steady intellectual ascent in which superior innovations replace outdated methods. It is, rather, the story of the development of a variety of techniques to the point where each is able to solve the problems currently of interest. In this it seems typical of the history of mathematics.

Exercises

1. Show that if two points on the surface of the sphere are not diametrically opposite then there is a unique great circle containing them, but that, when the two points are diametrically opposite, then there are infinitely many such great circles.
2. Show that the intersection of a plane with a sphere is either a single point or a circle on the sphere, and conversely.

3. Show that any parallel circle is parallel to a unique great circle.
4. Show that any two great circles on a given sphere bisect each other.
5. Suppose in the figure of the complete quadrilateral that arcs $\widehat{AG}$, $\widehat{GD}$ and $\widehat{AB}$ are quadrants, and that $\widehat{AE} = 60°$ and $\widehat{DZ} = 45°$. Use Menelaos' Theorem to calculate $\widehat{AD}$.
6. Use Menelaos' Theorem, and the fact that the angle ε between the equator and the ecliptic is approximately $23\frac{1}{2}°$, to find δ, the height of the sun above the equator, on the day when it is at a point on the ecliptic 45° south (measured along the ecliptic) of where the ecliptic and equator intersect. According to Ptolemy's Table of Chords, Crd 47° = 47; 51. (*Hint*: In Fig. 6.2 take BA the equator, BE the ecliptic, G the north pole and Z the sun.)
7. Let P be the vertex of a given cone and π and π' parallel planes that do not contain P. Then π intersects the cone in a circle if and only if π' does. Prove this.
8. Show that stereographic projection maps any circle parallel to the equator onto a circle.
9. In Fig. 6.5 show that $CD = (\cos(\varepsilon)\cdot NS)$ and that $C'D' = CD/[1 + \mathrm{Sin}(\varepsilon)]$. Use these results to construct the image, under stereographic projection, of the parallel circle of latitude δ.
10. In Fig. 6.5 take A′B′ as the X-axis and O as the origin. Show that if a great circle whose diameter is XY is inclined to the equator at an angle of θ then its stereographic image is a circle whose diameter is $\tan(45° - \theta/2)\tan(45° + \theta/2)$ and whose center is at the point:

$$\left(\tan\left(45° - \frac{\theta}{2}\right) - \frac{1}{2}\left[\tan\left(45° - \frac{\theta}{2}\right) + \tan\left(45° + \frac{\theta}{2}\right)\right], \theta\right).$$

11. With reference to Fig. 6.5 calculate the center and diameter of a circle whose diameter ZW is parallel to XY so that $\widehat{WY} = \beta$.
12. Use the results of the above problems to construct the lines on the plate of an astrolabe for your latitude that show the equator and the two tropics, your horizon, and almucantars for every 6°.
13. From a newspaper find the length of daylight for a particular day and then calculate the times, according to your clock, at which the (seasonal) hours of the night begin.
14. If you have access to an astrolabe try to calculate the length of the seasonal hours of the day without consulting a newspaper or other such aid. (You will have to know the date and will have to use the back of the astrolabe to find the position of the sun in the ecliptic for the given date.)
15. If you have access to an astrolabe use it to find the declination of the sun for a given date. Describe the method you use.
16. Use the Rule of Four Quantities and the data in Exercise 6 to compute the declination of the sun.
17. Show that Menelaos' Theorem implies the Rule of Four Quantities, which we know implies the Sine Theorem. Show that the Sine Theorem implies Menelaos' Theorem and conclude that the three results are equivalent.

18. Solve Exercise 6 using the Sine Theorem.
19. Show that if one knows the qibla for a locality then the Sine Theorem may be used to obtain the great circle distance between the locality and Mecca.
20. Prove (P1) and (P2) of Section 8 from the principle that X is on the horizon of Y if and only if Y is on the horizon of X and the Basic Facts of Section 1.

Bibliography

Berggren, J. L; "A Comparison of Four Analemmas for Determining the Azimuth of the Qibla." *Journal for the History of Arabic Science* **4** (1) (1980), 69–80.

Berggren, J. L. "Spherical Trigonometry in the *zīj* of Kūshyār ibn Labbān". In: *Festschrift: A Volume of Studies of the History of Science in the Near East, Dedicated to E. S. Kennedy* (D. A. King and G. Saliba, eds.) New York: New York Academy of Sciences (to appear).

An English translation of the section in the *zīj* dealing with spherical trigonometry.

Bīrūnī, Abu l-Rayḥān (transl. by J. Ali). *The Determination of the Coordinates of Cities.* Beirut: American University of Beirut, 1967.

Debarnot, M.-Th. "Introduction du Triangle Polaire par Abū Naṣr b. ʿIrāq." *Journal for the History of Arabic Science* **2** (1) (1978), 126–136.

Kennedy, E. S. *A Commentary Upon Bīrūnī's Kitāb Taḥdīd al-Amākin.* Beirut: American University of Beirut, 1973, (Commentary on Bīrūnī's *The Determination, etc.*)

King, D. A. "Ibn Yūnus' Very Useful Table for Reckoning Time from the Sun", *Archive for History of Exact Sciences* **10** (1973), 342–394.

The fundamental study of the Cairo Corpus of tables for astonomical timekeeping.

King, D. A. Art. "Ḳibla". In: *Encyclopedia of Islam*, 2nd ed. Leiden: E. J. Brill, 1979.

Lorch, R., "Al-Khāzinī's 'Sphere That Rotates by Itself'." *Journal for the History of Arabic Science* **4** (2) (1980), 287–329.

MacAlister, Paul R. and Flolydia M. Etting (designers). *The Astrolabe Kit* with *The Astrolabe: Some Notes, etc.* by R. S. Webster. Lake Bluff, IL: Paul MacAlister, 1974.

Index

This index refers to all preceding parts of the book other than the Contents, the Preface and the Exercises to each chapter. For the purposes of alphabetizing, the Arabic article "al-" has been counted as part of the Arabic name, so that, e.g. al-Manṣūr is listed under "A" and not under "M". However, the Arabic letter "ʿayn" (transliterated as "ʿ") has been ignored. Books are listed under the name of the author, but works listed only in the bibliographies have not been indexed. A word in quotation marks, e.g. "algebra", refers to information on the etymology of that word. The numbers after each entry refer to pages.